ArcGIS™ Developer's Guide for VBA

Amir H. Razavi

THOMSON

DELMAR LEARNING

Australia   Canada   Mexico   Singapore   Spain   United Kingdom   United States

ArcGIS™ Developer's Guide for VBA
Amir H. Razavi

Publisher:
Alar Elken

Executive Editor:
Sandy Clark

Acquisitions Editor:
James De Voe

Development Editor:
John Fisher

Executive Production Manager:
Mary Ellen Black

Executive Marketing Manager:
Maura Theriault

Marketing Coordinator:
Sarena Douglas

Channel Manager:
Fair Huntoon

Production Manager:
Larry Main

Production Editor:
Betsy Hough

Editorial:
Carol Leyba, Daril Bentley

Cover Design:
Cammi Noah

Trademarks
ArcGIS, ArcInfo, ArcMap, ArcObjects, and ArcView are registered trademarks of Environmental Systems Research Institute (ESRI).

For more information, contact:
Delmar Learning, Executive Woods, 5 Maxwell Drive, Clifton Park, NY 12065-2919. Or find us on the World Wide Web at
http://www.onwordpress.com

For permission to use material from this text or product, contact us by
Tel : 1-800-730-2214
Fax: 1-800-730-2215
www.thomsonrights.com

ISBN: 0-7668-6325-5

NOTICE TO THE READER

About the Author

Amir H. Razavi is a professional engineer registered in the states of Maryland and Virginia. He has a B.A. in civil engineering and a Master's of Information Systems Management from George Washington University. Amir has been developing software since 1982, and has served as GIS manager for the Civil Rights Division at the Department of Justice. He has also been involved in large-scale software development projects, such as the Agent Registration System of the National Association of Securities Dealers.

In 1994, Amir founded Razavi Application Developers (RAD), which specializes in implementing GIS and database systems, among others. He is coauthor of *ArcView/Avenue Programmer's Reference* (OnWord Press). The author can be reached at *arazavi@razavi.com*. See RAD's web site at *www.razavi.com*.

Acknowledgments

This book would not have been a reality without the efforts and guidance of the people at OnWord Press/Delmar Publishers. I am very grateful to them. Thanks also to Jim Peroutky of ESRI for his thoughtful and highly professional technical review of the manuscript.

I have been blessed with a wonderful family that has supported me in every endeavor. Thus, I am dedicating this book to everyone in my family: to my wife, Tema, who inspired me to write this book; to my son, Dean, who cheered me along the way; to my parents, who have dedicated their lives to their children; and to my sisters, Homeira and Marva.

Contents

Introduction

ARCGIS, DEVELOPED AND DISTRIBUTED BY Environmental Systems Research Institute (ESRI), is a scalable software system for creating, managing, and analyzing geographic data. It is a set of integrated GIS products that form a complete software package. ArcMap, ArcToolbox, and ArcCatalog are the core products of ArcGIS. ArcMap is used for creating, maintaining, and analyzing maps. ArcToolbox provides for data conversion and geoprocessing tasks. ArcCatalog manages spatial data.

The technology framework of ArcGIS is known as ArcObjects. You can customize and extend the capabilities of ArcGIS using the ArcObjects technology. This technology is based on Microsoft's Component Object Model (COM) specification. Therefore, you can use any programming language (such as Visual Basic or C + +) that supports COM in developing ArcGIS applications.

This book focuses on ArcMap's development and customization applications using the Visual Basic for Applications (VBA) platform. The book's goal is to teach you the basics of developing ArcGIS applications; it does not show you everything in ArcObjects. That would require multiple volumes and thousands of pages. The knowledge you gain by reading this book helps you in going beyond the basics toward developing complex ArcGIS applications.

Audience

This book is for professionals, instructors, and advanced students who wish to customize or develop ArcGIS applications using VBA. You do not have to be a seasoned programmer to understand this book. However, this book is about programming ArcGIS with VBA, and is not a programming tutorial. You must have some familiarity with a programming language, preferably VBA or Visual Basic. You also do not have to be a GIS expert to benefit from this book, but effective application developers have a good understanding of the target environment. You should have some basic familiarity with ArcGIS.

Content

This book covers the VBA, COM, and ArcObjects technologies as necessary to achieve its objectives. Each of these might be the subject of an individual book explored in detail. The basic concepts (and details) of these technologies are covered in this book to the degree described in the following overview of chapter content.

Chapters 1 through 3 discuss the development environment and programming language of VBA. If you have used VBA extensively you can skip Chapter 1, but review chapters 2 and 3, in that these chapters contain important information on the integration of VBA and ArcObjects.

Chapter 4 reviews your options for distributing your application and how you can protect your code. Chapters 5 and 6 discuss the COM technology. COM is a specification for creating software components that are independent of programming languages. Because ArcObjects is based on the COM specification, an understanding of the COM technology helps you become more efficient.

Chapter 7 covers topics related to developing ArcObjects applications and macros with COM. It shows you how to start every ArcGIS application, and offers a set of programming guidelines for ArcObjects applications.

Chapters 8 and 9 discuss ArcObjects basics and elements. In these chapters you learn how to read the ArcObjects object model diagram. You are also presented with a methodology for developing ArcGIS applications.

At the end of Chapter 9 you will have learned about VBA, COM, and ArcObjects, and will be ready to begin developing ArcGIS applications. Chapter 10 is a quick-start tutorial showing you how a typical ArcGIS application is developed.

Chapter 11 concentrates on the customization aspect of developing applications. In this chapter you learn how to customize ArcMap and to create a user interface for your application.

Chapter 12 reviews the model on which the objects used in chapters 13 through 16 are based. You learn how to develop applications using specific elements of ArcMap in chapters 13 through 16.

Typographical Conventions

The following are the typographical conventions used in this book. These include the following uses of icons for easy location of notes, tips, and attendant web site material. Information is also provided regarding the distinguishing characteristics of various text items.

TIP: *Tips, which appear in this form, are the fruit of experience and are aimed at saving you time and stress.*

Scripts following this symbol are available at the address provided at the end of this introduction, under "Book Scripts and Downloadable Web Script Files."

The following are the distinguishing characteristics of various text items.

• Examples of VBA macros are set off from the text and appear in a monospaced typeface, an example of which follows.

```
{This was an Avenue example]
Set pMxDocument = Application.Document
```

• Menu items, tool buttons, and programming elements (such as the names of requests and object classes) are capitalized.

• User input; the names of files, directories, and fields; and script variables and other script elements appear in the text proper in italic.

Book Scripts and Downloadable Web Script Files

Complete program code files for the procedures shown and referenced in this book are available on the Web. The web icon next to the start of code examples bears the location information for the code file, such as *CODE VBA3-1* (the file on the Web bears the same file name). To access the code files, go to the following address and follow the instructions for downloading.

http://www.onwordpress.com/resources/olcs/razavi/arc_gis_vba

CHAPTER 1

The VBA Development Environment

VISUAL BASIC FOR APPLICATIONS (VBA) is one of the best technologies developed in recent years. It is a programming language developed by Microsoft. Applications such as ArcGIS that expose their Component Object Model can integrate with VBA. You will learn more about the Component Object Model later in this book. There are many other applications that use VBA, such as Microsoft Word and Access. VBA has standardized the development of custom applications for these Windows-based products. Visual Basic is one of the easier programming languages to learn.

The applications developed using VBA are often called macros. You can use macros to automate repetitive tasks or to create complete applications. The macros run inside the applications that house VBA. For example, your ArcMap macros run inside the ArcMap application and act on the ArcMap document. This chapter describes the VBA development environment, which is the VBA editor. You use the VBA editor to develop your macros. You can also use your VBA editor to test and debug your macros. Yet another use of VBA is to prototype larger applications. However, keep in mind that VBA code may not easily translate into a Visual Basic application, especially if you use forms.

Opening and Closing the VBA Editor

Your VBA macro is stored with the ArcGIS document. Therefore, to access the VBA macro you need to first open an ArcGIS document, such as an ArcMap document. You start the ArcMap application with a new or existing document, and you add or edit the VBA macro using the VBA editor. To open the VBA editor, you select Tools | Macros | Visual Basic Editor. Alternatively, you can use the shortcut key combination Alt + F11. The VBA editor will open in a new window. Figure 1-1 shows you how the VBA editor might appear when opened in a new ArcMap document.

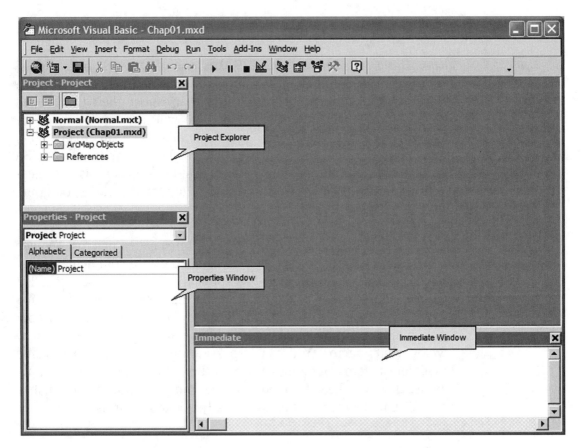

Fig. 1-1. The VBA editor.

The Project Explorer window in the VBA editor lists the various components of your macro. In a new ArcMap document, the VBA editor gives you access to the Normal template and the current project document. If you place your macro in the Normal template, it becomes available to all of your documents. When you place your macro in the Project document, it is only available in that document.

You view and edit properties of an object in the Properties window. In figure 1-1, the Project element is highlighted in the Project Explorer window. Therefore, the Project element's property is shown in the Properties window. In this case, the Project has only one property, known as Name. You can change the Name property by providing a new name in the Properties window. The Immediate window is used when testing and debugging your macro. You will learn more about how to use this window in later chapters.

You close the VBA editor by selecting File | Close and Return to Arc-Map. Alternatively, you can use the shortcut key combination Alt + Q, or click on the system's Close button at the upper right corner of the window. When you save the ArcMap document from the ArcMap window or VBA editor, all changes, including macros, are saved. Therefore, if you close the VBA editor before saving your new macro, the macro is not lost; you can save it from the ArcMap window. When you exit Arc-Map, all related windows, including the VBA editor, are closed.

The VBA Editor User Interface

The VBA editor user interface consists of a menu bar, toolbars, and several windows. You can customize this user interface by moving, hiding, or displaying any of the user elements. Figure 1-2 shows the menu bar and standard toolbar. You cannot hide the menu bar, but you can move it around.

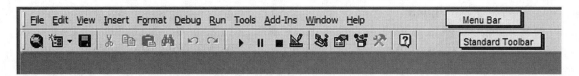

Fig. 1-2. The VBA editor menu and toolbar.

The VBA editor offers four other toolbars. You can display or hide the toolbars via the VBA editor's Customize dialog box. Select View | Toolbars | Customize to display the Customize dialog box, shown in figure 1-3. A check mark next to a toolbar on the Toolbars tab displays that toolbar.

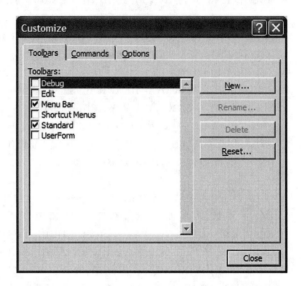

Fig. 1-3. The VBA editor Customize dialog box.

Similar to other Windows applications, the VBA editor offers a complete set of menu options. The most commonly used and important menu items are discussed in the sections that follow.

File Menu

The Save and Close items in the File menu are most often used to save the document or close the editor. In this menu option, the Import File and Export File items are noteworthy. The program codes you develop for your macro are stored in the current document or in the Normal template. Program codes are not accessible outside the document or template.

You can select File | Export File to save your macro in a text file outside the ArcGIS document. The exported file retains the format of Visual Basic files. You can use the file in a Visual Basic project. Exporting your macro is also a good way to make backups or share your code with others. Conversely, you can import Visual Basic files into your document by selecting File | Import File.

View Menu

Use the items in the View menu to display various VBA editor windows. For example, if you accidentally close the Project Explorer window, you can display it again by selecting View | Project Explorer. This menu also provides for switching to the ArcMap or ArcCatalog window.

Insert Menu

You will use the items in the Insert menu to add various elements to your macro. For example, you can add user forms to your macro by selecting Insert | User Form.

Debug and Run Menus

The items in the Debug and Run menus allow you to run, test, and debug your macro. Selecting Debug | Step, you can walk through your macro one statement at a time. This is a very effective and common method of debugging a macro.

Using the VBA Editor

You enter and maintain the programming code that constitutes your macro in the VBA editor's Code window. There are several ways of opening and displaying the Code window. One is to use View | Code. When you insert a module or class module via the Insert menu option, the Code window for the new module automatically opens. The steps that follow take you through the process of creating a new macro that will also open the Code window.

1   Start ArcMap and open the VBA editor by pressing Alt + F11.

2   In the VBA editor window, select Tools | Macros to display the Macros dialog box, shown in figure 1-4.

3   Ensure that the value in the Macros In drop-down list is set to your current project. Then, enter *MyMacro* in the Macro Name text box.

4 Click on the Create button to open the Code window, shown in figure 1-5.

Fig. 1-4. The VBA editor's Macros dialog box.

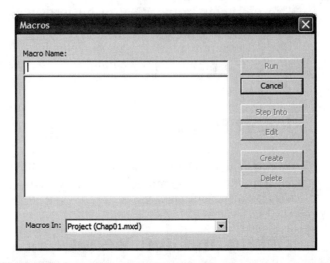

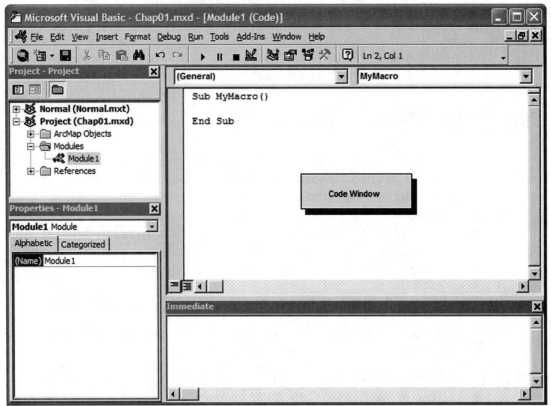

Fig. 1-5. The Code window.

In figure 1-5, the Code window has the framework for building a macro. All you need to do is add code. Add the *MsgBox* statement in the following code segment to your macro.

```
Sub MyMacro()
  MsgBox ("Hello World!")
End Sub
```

Run this new macro by selecting Run | Run Sub. The result should be a message box containing the text *Hello world!* In figure 1-5, you can see that a new element named *Module1* has been added to the Project Explorer window. The macro you just created is a procedure in *Module1*. You can change the name *Module1* in the Properties window to a more meaningful name. You can have more than one macro in a module. You can also have multiple modules in a project.

Creating User Forms in the VBA Editor

User forms offer a rich set of user interface controls you can add to your macro. Although there are several built-in dialog boxes you can use in your macro to interact with the user, complex macros may need their own custom forms. You add forms to your macro by selecting Insert | User Form. Figure 1-6 shows a newly inserted user form.

In figure 1-6, the new user form is named *UserForm1*. It is also added to the Project Explorer window. You can rename the form in the Properties window. You can add user interface controls to your form via the Toolbox dialog box. If the Toolbox dialog box is not shown, select View | Toolbox to display it. In the following steps, you will add a text box and a command button to the form. Then you will add code so that clicking on the form's command button displays the current date and time in the text box.

1   Start ArcMap and open the VBA editor.

2   In the Project Explorer window, select the current Project document. Do not select the Normal template.

3   Select Insert | User Form to insert a new form in the current project document.

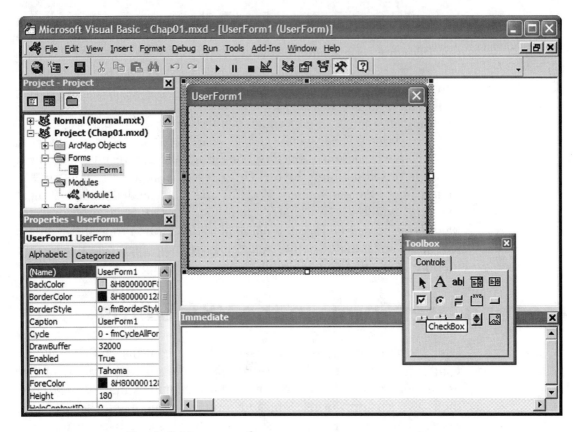

Fig. 1-6. New user form.

4   If the Toolbox dialog is hidden, select View | Toolbox to display it. In the Toolbox dialog, select the Command Button tool. If you are not familiar with the tool icons, place the mouse pointer on a tool to view its name.

5   While the Command Button tool is selected, click on the user form to add a command button.

6   Select the Text Box tool from the Toolbox dialog. Click on the user form to place a text box on the form.

7   Arrange the user interface controls by dragging or resizing them so that the form appears as shown in figure 1-7.

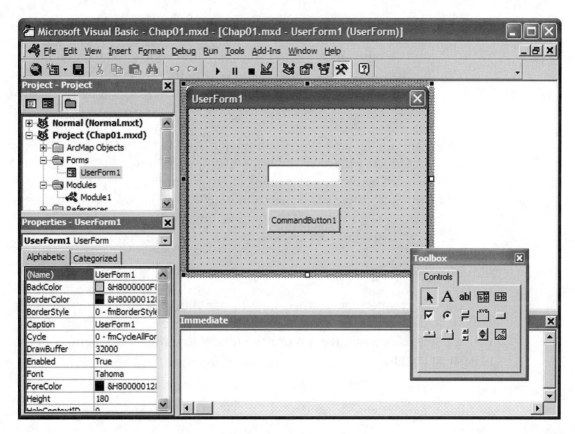

Fig. 1-7. User interface controls.

8 Add the code for displaying the date and time. Right-click on *CommandButton1* and select the View Code pop-up menu item. This action displays the Code window showing the following statements.

```
Private Sub CommandButton1_Click()
End Sub
```

9 Add the statement *Me.TextBox1.Text = Now()* so that the Code window contains the following procedure.

```
Private Sub CommandButton1_Click()
  Me.TextBox1.Text = Now()
End Sub
```

10   Test your form by selecting Run | Run Sub | UserForm while your cursor is inside the *CommandButton1_Click()* procedure. The user form should appear. When you click on the form's command button, you should see the date and time in the text box. Close the user form by clicking on the Close button at the upper right corner of the form.

You can run macros directly from the ArcMap window by selecting Tools | Macros | Macros. However, user forms do not appear in the list of macros. Therefore, you cannot run user forms directly from the Arc-Map window. You need a macro that displays your user form. Create a new macro as you learned in the previous section, and add the following code to display your user form.

```
UserForm1.Show
```

In this chapter you learned how to use the VBA editor to create new macros. Macros are written in the Visual Basic for Applications (VBA) language. You will learn more about how to write VBA programs in the next chapter.

CHAPTER 2

The VBA Programming Language

TO BECOME PROFICIENT IN DEVELOPING ArcObjects macros and applications, you need to first become fluent in the VBA programming language. This chapter presents VBA's basic programming elements, which include variables and control statements. If you have studied other programming languages, the VBA language will be familiar to you. Even if you are not a programmer, VBA is easy to learn.

Referencing Objects with Variables

Your application can guide ArcGIS to execute many types of tasks, such as loading a shape file or preparing a map layout. To accomplish these and other tasks, your VBA program needs information such as the shape file name or page size for layout. These types of information are stored in variables.

A variable is an area in memory reserved for storing a piece of information that can be accessed or changed by a program. Variables can store data types such as text strings and numbers. They can also reference objects. Objects are the things you work with when developing ArcGIS applications. For example, when an application turns on the visibility of a layer, it needs to reference the *Layer* object.

In ArcGIS, your application does not directly reference an object. Instead, it points to one of the object's interfaces. An object's interface organizes related properties and methods of an object. Throughout this book you will see many examples of how variables can point to interfaces to access objects in ArcObjects.

VBA requires that you declare variables in advance of their use. However, you can declare variables explicitly or implicitly. For implicit declaration, just use the variable. When VBA encounters the variable for the first time, it will declare it. In the case of explicit declaration, you declare the variable before using it. To declare a variable, you use one of the following keywords: *Dim*, *Private*, *Public*, or *Static*. For example, to declare the *MyVar* variable you use the following statement.

```
Dim MyVar
```

The *Dim* keyword is the most used method of declaration. It declares a variable to be used inside a procedure. Other keywords for declaration have special purposes. For example, declaring with the *Public* keyword makes the variable available to more than one procedure. The preceding example of a declaration creates a variable named *MyVar* as a variant. Variant variables can accept different types of data. You can explicitly indicate the data type a variable can accept, as in the following.

```
Dim MyVar As Long
```

In the preceding example *MyVar* can only accept numeric data types of long integers. The following are other common data types.

- *Boolean:* Variables of this type can be either True or False.
- *Date:* Date and time values are stored in these types of variables.
- *Double:* Variables of this type store numerical values with decimal point.
- *String:* This data type is for storing text.
- *Object:* Variables of this type point to objects.

You are strongly urged to always declare variables. You can force the explicit declaration by adding the following line to the beginning of your module.

```
Option Explicit
```

You can declare the variable anywhere in your macro, as long as it is before its first use. You can also declare multiple variables on a single *Dim* statement, separating each variable with a comma.

Using the Global Application Objects

VBA offers two objects that are always available to you while running ArcMap. The two objects can be accessed with the keywords *Application* and *ThisDocument*. The *Application* keyword references the Arc-Map program, and *ThisDocument* points to the current ArcMap document. The following example shows a macro that displays the name of the first layer in a map. The macro produces an error if there are no layers.

```
Option Explicit
Sub MyMacro()
   Dim pMxDocument As IMxDocument
   Dim pMap As IMap
   Dim pLayer As ILayer
   Set pMxDocument = Application.Document
   Set pMap = pMxDocument.FocusMap
   Set pLayer = pMap.Layer(0)
   MsgBox pLayer.Name
End Sub
```

In the preceding example there are three variable declarations. Each variable is declared as an ArcObjects interface. In this manner they point to an object through the object's interface. The *Set pMxDocument = Application.Document* statement points the *pMxDocument* to the *IMxDocument* interface to access the ArcMap document. The *Document* property of the *Application* global object returns the pointer to the interface.

Scopes of the Variable

Scope of a variable refers to the area of VBA where the variable is available. A variable declared in a procedure is not available outside that procedure. Such a variable is often known as a local variable. A variable can be declared outside the procedures of a module so that it is available to all procedures of that module.

You can declare local variables using the *Dim* or *Static* keyword inside a procedure. The macro in the last section has an example of declaring three local variables. The following example shows you how to declare variables outside the procedure.

```
Option Explicit
Public pMxDocument As IMxDocument
Private pMap As IMap
Sub MyMacro()
   Dim pLayer As ILayer
   Set pMxDocument = Application.Document
   Set pMap = pMxDocument.FocusMap
   Set pLayer = pMap.Layer(0)
   MsgBox pLayer.Name
End Sub
```

In the preceding example the *pMxDocument* variable is declared with the *Public* keyword. A variable with the *Public* scope is available to all procedures in all modules. The *Private* scope of the *pMap* variable makes it available to all procedures of the module making the declaration.

Writing VBA Statements

You develop VBA macros by writing VBA statements. Among statement types, *assignment* is the most common. The *assignment* statement consists of an equals sign with a variable on its left and an expression or object to its right. The result of the expression or the object on the right is then assigned to the variable on the left. The following are two *assignment* statements.

```
SName = "Alaska"
Set pMxDocument = Application.Document
```

You need to use the *Set* keyword when assigning an object. The following sections discuss other statement types.

Conditional Statement

Controlling the flow of logic in any programming language is a basic operation. Such controls range from executing a set of statements if a certain condition prevails to executing the same set more than once.

An *If* statement is used to conditionally execute a series of statements. The condition is a Boolean expression resulting in *True* or *False*. For instance, you may want to make a layer visible if its name is *"STATE"*, as shown in the following code segment.

```
If pLayer.Name = "STATE" Then
  pLayer.Visible = True
End If
```

The condition is *pLayer.Name = "STATE"*. If the layer name is *"STATE"*, the condition returns *True*; otherwise, it returns *False*. When the condition is *True*, the statements between the *If* and *End If* lines are executed. *If* statements can also be nested. A nested *If* structure is a conditional block inside another *If* block. An *If* structure starts with the following statement.

```
If condition Then
```

The structure can have the following optional statements.

```
ElseIf condition Then
```

or

```
Else
```

An *If* structure must end with the following statement.

```
End If
```

VBA executes the program lines following the *If* statement if the condition expression is true. Otherwise, the execution moves to the *ElseIf* or *Else* lines if provided.

Loop Structure

Loop structures execute a set of code lines more than once. For example, you may write one set of code to edit a given layer. Then you iterate through all layers with a loop structure, making the same edits. There are two types of loops. Fixed iteration loops repeat for a predefined number of times. Indefinite loops repeat until a condition stops the loop.

Fixed iteration loops are used when the numbers of iterations are known in advance. For example, you may want to loop through all layers of a map and make each visible. The *For/Next* loop structure is used for fixed iterations. The syntax follows.

```
For counter = start To end
    ...
Next counter
```

The following example iterates through the layers of the active map.

```
For LayerCount = 0 To pMxDocument.FocusMap.LayerCount - 1
    ...
Next LayerCount
```

VBA starts the loop by assigning the start value to the counter. It executes the statements inside the loop structure. When VBA reaches the *Next* statement, it increments the counter and repeats the loop until the counter reaches the end value.

The *Do/While* loop structure is used for indefinite loops. Such loops are used when the number of iterations is not known in advance. The following code segment shows you an example of looping through features of a layer.

```
Set pEnumFeature = pMxDocument.FocusMap.FeatureSelection
pEnumFeature.Reset
Set pFeature = pEnumFeature.Next
Do While Not pFeature Is Nothing
    ...
    Set pFeature = pEnumFeature.Next
Loop
```

In the preceding example the loop repeats as long as the variable *pFeature* is not "Nothing." *Nothing* is a special value for object variables without an object. When *pEnumFeature.Next* runs out of features, the value of *pFeature* is set to *Nothing*. You must be very careful with these types of loops. Unless you account for a way to end the loop, you could have an infinite loop that would run forever. In the preceding example, the *Set pFeature* = *pEnumFeature.Next* statement inside the loop structure will eventually cause the loop to end.

Continuation Statement

When you write VBA programs, each statement must be placed on one line. You can break a statement into multiple lines by adding the continuation character to the end of the incomplete lines. The continuation character is the underscore. The following code segment shows how a continuation character can break a statement into two lines.

```
Set pFeature = _
pEnumFeature.Next
```

Because of the page width limit in this book, many of the VBA statements are broken into multiple lines using the continuation character.

Adding Comments

Documenting your program can save time in the future when you need to maintain the application code. The VBA character for comments is a single quote ('). Whenever this character appears, everything to the end of the line is read as a comment. The exception is when a single quote appears inside a text string.

Creating Procedures

The VBA code you write must be placed inside procedures. A macro can consist of one or more procedures. There are two types of procedures: subs and functions. The primary difference between the two is that a function can return a value. A procedure starts with the keyword *Sub* or *Function* and ends with the *End Sub* or *End Function* statement.

VBA automatically adds the starting and ending statements when you use the Macro dialog or Insert menu option to add new procedures.

You can pass variables between procedures by using arguments. The following code segment shows an example of the *Function* statement.

```
Function CalculateSquareRoot(NumberArg As Double) As Double
  ...
End Function
```

VBA has many built-in functions you can use in your macro. The following are common built-in functions that manipulate text strings.

- *InStr (start, string1, string2, compare):* Returns the position of the first instance of the *string2* inside *string1*.

- *Len (string):* Returns the length of the string.

- *Mid (string, start, length):* Returns parts of the string that start at the start position for the given length.

The following code segment displays the text *Maryland* in a message box.

```
Public Sub MySub()
  Dim strName As String
  strName = "Maryland"
  MsgBox Mid(strName, 5, 4)
End Sub
```

CHAPTER **3**

Debugging and Error Handling

VBA STOPS RUNNING AND DISPLAYS AN ERROR MESSAGE when it encounters an error in your macro. You should thoroughly test your macro before relying on it in your work. An error that causes your macro to execute partially or incorrectly is known as a bug. Debugging is the process of testing your macro and removing every bug.

Error handlers are the sections of your macro that trap errors and take specific action. You should include error-handling code in your large or complicated macros. As your macro becomes larger and more complicated, it becomes more difficult and costlier to test and debug for every possible situation. Error handlers are like an insurance policy that if an error does occur you could trap it. Macros that are less than a dozen statements may not need error handlers. In this chapter you learn how to test and debug your macro. You also learn how to place error handlers in your macro.

Types of Errors

You should watch for the following there types of errors that may exist in your macro. These are discussed in the sections that follow.

- Compile or language error
- Runtime error
- Logic error

Compile or Language Error

VBA is compiled before running. Therefore, if you have made a mistake in the VBA programming syntax, it would not compile. If you tried to compile the following macro, you would get the errors shown in figure 3-1. You can compile automatically by running the macro or by selecting Debug | Compile Project. Once the error message is displayed, the *MxDocument* is highlighted in your code.

```
Sub MyMacro()
   Dim pMxDocument As IMxDocument
   pMxDocument = Application.Document
   MsgBox pMxDocument.FocusMap.Name
End Sub
```

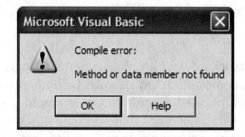

Fig. 3-1.
A compile error.

VBA highlights the *pMxDocument* variable, indicating that it is causing the error. The problem is that this is a variable for storing an object. Therefore, you should use the *Set* keyword for its assignment. The following macro corrects this problem.

CODE
VBA03-1

```
Sub MyMacro()
   Dim pMxDocument As IMxDocument
   Set pMxDocument = Application.Document
   MsgBox pMxDocument.FocusMap.Name
End Sub
```

Runtime Error

A runtime error, as its name implies, occurs while the program is running. In these cases, the programming syntax is correct but something else has gone wrong. Consider the following macro. In this macro the name of the second map (also known as *dataset in* ArcMap) is displayed. However, if there is only one map in the ArcMap document, the error shown in figure 3-2 is displayed when running the macro.

```
Sub MyMacro()
   Dim pMxDocument As IMxDocument
   Dim pMaps As IMaps
   Dim pMap As IMap
   Set pMxDocument = Application.Document
   Set pMaps = pMxDocument.Maps
   Set pMap = pMaps.Item(1)
   MsgBox pMap.Name
End Sub
```

Fig. 3-2.
A runtime error.

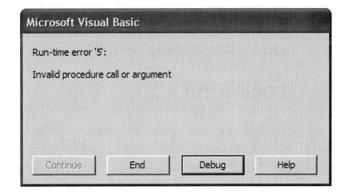

The preceding macro compiles without a problem, but produces the error message shown in figure 3-2 when there is only one map in the document. Selecting the End button in the Error dialog interrupts the running of the macro. Selecting the Debug button places the macro in break mode and highlights the offending code line. A break mode in VBA is when execution of a program is temporarily suspended. In the preceding example, when you click on the Debug button, VBA highlights the following line.

```
Set pMap = pMaps.Item(1)
```

The preceding statement attempts to get the second *Map* object from the Maps collection. The first map would be *Item(0)*. A better programming approach would be to test the number of maps in the document, as shown in the following macro.

```
Sub MyMacro()
  Dim pMxDocument As IMxDocument
  Dim pMaps As IMaps
  Dim pMap As IMap
  Set pMxDocument = Application.Document
  Set pMaps = pMxDocument.Maps
  If pMaps.Count > 1 Then
    Set pMap = pMaps.Item(1)
    MsgBox pMap.Name
  End If
End Sub
```

Another source of runtime error is mistyping variable names. You can eliminate this possibility by adding the *Option Explicit* statement to the beginning of your module. This statement forces you to declare each variable, whereby a mistyped variable is never declared and becomes a compile error that can be fixed easier than a runtime error.

Logic Error

Logic errors are generally toughest to identify and correct. A logic error may not interrupt the execution of your macro. Instead, it just produces an erroneous result. The following macro demonstrates a simple logic error. This macro counts the number of feature layers in the active map. However, it always displays a number that is one more than the correct value.

```
Sub MyMacro()
  Dim pMxDocument As IMxDocument
  Dim pMap As IMap
  Dim lCount As Long
  Dim lIndex As Long
  Set pMxDocument = Application.Document
  Set pMap = pMxDocument.FocusMap
  lCount = 1
  For lIndex = 0 To (pMap.LayerCount - 1)
    If TypeOf pMap.Layer(lIndex) Is _
    IFeatureLayer Then
      lCount = lCount + 1
    End If
```

```
   Next lIndex
   MsgBox "Number of the feature layers " & _
   "in the active map: " & lCount
End Sub
```

The best way to correct logic errors is to walk through the code and examine the result of each statement. You will learn how to walk through the code and debug the preceding macro later in this chapter.

Debugging Compile and Runtime Errors

Compile and runtime errors interrupt the execution of the macro and highlight the statements containing errors. To correct compile errors, you need to know the correct language syntax. The VBA editor also helps you avoid some of the compile errors while typing code. If you type a statement with an obvious error, such as an *If* statement without the *Then* keyword, the VBA editor displays an error message when you go to the next line. If you do not want to see any error messages while typing code into the VBA editor, deselect Auto Syntax Check in the Tools | Options dialog.

VBA gives you the option of going into break mode when a runtime error has occurred. You select the Debug button in the Error dialog to enter break mode. In break mode, the VBA editor temporarily suspends the macro's execution at the statement containing the error. In break mode, you can use the VBA editor's Immediate window to investigate the problem.

You can open the Immediate window by selecting View | Immediate Window. This widow is like a virtual notepad in which you can execute VBA statements or examine values of variables in your macro. However, in the Immediate window you cannot declare variables and you cannot execute multi-line statements such as a loop or *If* structure.

You can print from your code onto the Immediate window by using the *Debug.Print* statement. The following statement prints the value of variable *lCount* in the Immediate window.

```
Debug.Print lCount
```

Using the *Debug.Print* statement shows the values of variables or expressions in the Immediate window of VBA as the macro is running. You can also use the *?* operator while in break mode to get variable values. If you type the following statement in the Immediate window while in break mode, you get the value of the *lCount* variable.

```
? lCount
```

Many of the runtime errors for ArcMap macros are caused by the fact that ArcObjects cannot return a valid object or interface due to an earlier problem in your code. In such cases your objects are set to *Nothing*. *Nothing* is a special value that indicates your variable is not associated with any object. You can test in the Immediate window if a variable is pointing to an object. Enter the following statement in the Immediate window while in break mode.

```
? pMap is Nothing
```

Assuming you have declared a variable named *pMap* in your macro, the Immediate window returns *True* if *pMap* is *Nothing* (not associated with any object).

Debugging Logic Errors

An effective approach to debugging logic errors is to walk through the code. Walking through the code means that you execute your macro in break mode one statement at a time. After each statement, you can examine variable values or even change them to identify a logic error.

You can walk through your code from the beginning or from a specific statement. To start walking through your code, in the VBA editor, select Debug | Step Into. To go into break mode at a specific line, place a breakpoint on that line. When a breakpoint is placed on a statement, the macro stops and goes into break mode at that statement. Select Debug | Toggle Breakpoint to add or remove a breakpoint. The VBA editor highlights the next statement that will be executed when in break mode. Select Debug | Step Into (or press the F8 key) to execute the next statement. Pressing F8 walks you through the code.

As you walk through the code, you can use the Immediate window to examine variables' values. Earlier, the following macro was presented as having a logic error. The error is that the count of feature layers is wrong.

```
Sub MyMacro()
  Dim pMxDocument As IMxDocument
  Dim pMap As IMap
  Dim lCount As Long
  Dim lIndex As Long
  Set pMxDocument = Application.Document
  Set pMap = pMxDocument.FocusMap
  lCount = 1
  For lIndex = 0 To (pMap.LayerCount - 1)
    If TypeOf pMap.Layer(lIndex) Is _
    IFeatureLayer Then
      lCount = lCount + 1
    End If
  Next lIndex
  MsgBox "Number of the feature layers " & _
  "in the active map: " & lCount
End Sub
```

Because the error is in the *lCount* variable, place a breakpoint on the *lCount = lCount + 1* statement. Figure 3-3 shows how your Code window would appear with the breakpoint.

Fig. 3-3.
A breakpoint.

Run the macro by pressing the F5 key. The macro will start running and will stop at the statement with the breakpoint. At this point, the *lCount = lCount + 1* statement has not been executed. Examine the value of *lCount* by entering the following line in the Immediate window.

```
? lCount
```

The statement returns the value of 1. Press the F8 key to execute the highlighted statement and examine the value of *lCount* again. This time you notice that the value is 2. Although this is the first feature layer encountered, the count is already 2. Therefore, the problem is with the starting *lCount* value of 1; that is, the initial value of *lCount* before getting into the loop should be 0. You can now fix the problem by changing the *lCount = 1* statement to *lCount = 0*. The following macro reflects this correction.

CODE
VBA03-3

```
Sub MyMacro()
   Dim pMxDocument As IMxDocument
   Dim pMap As IMap
   Dim lCount As Long
   Dim lIndex As Long
   Set pMxDocument = Application.Document
   Set pMap = pMxDocument.FocusMap
   lCount = 0
   For lIndex = 0 To (pMap.LayerCount - 1)
     If TypeOf pMap.Layer(lIndex) Is _
     IFeatureLayer Then
       lCount = lCount + 1
     End If
   Next lIndex
   MsgBox "Number of the feature layers " & _
   "in the active map: " & lCount
End Sub
```

Another common logic error that causes the application to "hang" is the infinite loop. In ArcObjects many objects are accessed through a collection object. Generally you iterate through the members of these collections using a loop structure. If you forget to move to the next member of the collection inside the loop, you could get stuck with an infinite loop. If this happens, use the Ctrl + Break key combination to interrupt the execution of the macro.

Adding Error Handlers

The possibility of a runtime error always exists, especially if you are developing a large or complicated macro. The purpose of error handlers is first to prevent VBA from stopping when an error is detected. A second purpose is to provide your macro the opportunity to deal with the error.

Error handlers trap and process errors. You use the *On Error* statement to trap an error. The following macro has a generic error handler that catches all possible errors, reports them, and then ends the macro.

CODE
VBA03-4

```
Sub MyMacro()
    Dim pMxDocument As IMxDocument
    Dim pMaps As IMaps
    Dim pMap As IMap
    On Error GoTo SUB_ERROR
    Set pMxDocument = Application.Document
    Set pMaps = pMxDocument.Maps
    Set pMap = pMaps.Item(1)
    MsgBox pMap.Name
    Exit Sub
SUB_ERROR:
    MsgBox "Error: " & Err.Number & "-" & _
    Err.Description
End Sub
```

In the preceding macro, the following statements provide the error-handling process.

```
    On Error GoTo SUB_ERROR
...

    Exit Sub
SUB_ERROR:
    MsgBox "Error: " & Err.Number & "-" & _
    Err.Description
End Sub
```

The *On Error* statement traps the errors and redirects the program execution to the line labeled *SUB_ERROR*. This statement is generally placed at the beginning of the macro. If there are no errors, the *Exit Sub* statement ends the process after the message box is displayed. Without this statement, the execution continues and the macro displays the error message box when there is no error.

VBA executes the statements after the *SUB_ERROR* label when it encounters an error. In this case, the result is a message box displaying an error number and description. You can build on this generic error handler by examining the error number and taking appropriate action. The *On Error* statement has the following options in addition to *GoTo*.

- *On Error GoTo 0:* Disables the error trapping.

- *On Error Resume Next:* Continues the execution at the line after the offending statement.

You can override all error-handling statements by setting the Break on All Errors option of the Options dialog. You access the Options dialog by selecting Tools | Options. When this option is set, VBA goes into break mode when it encounters an error, regardless of your error handlers.

When there is an error, the *Err* object is populated with the appropriate information. Using the number and description of this object, you can display an error message. When a statement is executed without an error, *Err.Number* is set to zero. Therefore, you must access the error number and description immediately after the error has occurred.

CHAPTER 4

Distributing Your Application

VBA *MACROS ARE BEST SUITED FOR AUTOMATING* repetitive tasks or complicated processes. However, you can also develop complete applications using VBA. You may develop a VBA application for use in a particular project, or to be used with every project. Whether you are the only user of the application or you plan to provide your application to others, you have several options to consider. This chapter reviews the options available for distributing your application.

Selecting a Development Environment

VBA macros and applications are stored in an ArcGIS document or template. They do not exist outside the document or template. Consequently, if you plan to distribute your application independently of an ArcGIS document you should consider one of the other development environments. For example, you should consider Visual Basic or Visual C++ if you plan to develop an extension.

Storing the Application Code

If your VBA application is for a single project, you can store it in the related ArcMap document. It cannot be accessed by other documents. However, you can always export the code if you want to use it for another project.

You can store the VBA application in a template in order to include it in new documents. If you store it in the Normal template, all new documents will have the VBA application. You can create a new template and store the VBA code. Then, when selecting File | New, you can create a new document based on your own template (see figure 4-1).

Fig. 4-1. New document based on a template.

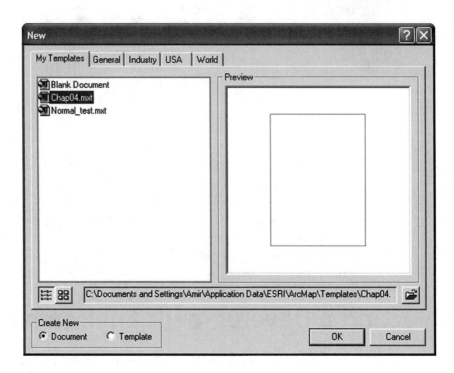

The following steps take you through the process of creating a template that includes your VBA application.

1 Start a new document and add your VBA application to the Project element in the VBA editor.

2 Select File | Save As to open the Save As dialog.

3 Select ArcMap Templates as the Save As Type option.

4 Provide a file name and click on the Save button.

ArcMap will list your template in the New dialog if you store it in the same directory as the Normal template. You can find out the Normal template's directory when you select File | New and the My Templates tab. In ArcCatalog, only the Normal template appears in the Project Explorer of the VBA editor. You cannot open ArcCatalog using a template, and it does not have a Project element. You can only add VBA macros to the Normal template.

Protecting Your Code

You can stop others from viewing and editing your VBA application code by locking the Normal or Project element in the VBA editor. Once you have locked the code, a password is required to unlock it. The following steps take you through the process of locking your VBA application.

1   In the VBA editor, right-click on the Normal template, Project template, or Project element in the Project Explorer window.

2   From the pop-up menu, select Properties, and then select the Protection tab from the Properties dialog.

3   In the Properties dialog, place a check mark next to *Lock project for* · *viewing* and provide a password. Save the document.

The next time you open the document and try to list the VBA code, you must enter the password. Remove the check mark (see previous step 3) to remove the lock.

CHAPTER 5

Component Object Model Basics

ARCOBJECTS IS BASED ON MICROSOFT'S COMPONENT OBJECT MODEL (COM) technology. COM is a specification for creating software components that are independent of programming languages or their location. When you develop ArcObjects applications with VBA, you use ArcObjects components. Therefore, an understanding of the COM technology can help you become more proficient with ArcObjects.

In this and next two chapters, you will learn about the COM technology and how to use it with ArcObjects applications. A complete discussion of COM is outside the scope of this book. However, the book covers the concepts that as a VBA programmer you should know.

Defining COM, ActiveX, and OLE

You should understand the difference between an application and a component. You use an application directly to carry out a task. For example, ArcMap is an application you can use to perform a spatial query. ArcMap is also a component because it offers an application programming interface (API) so that you can write your own application for performing a spatial query.

ArcMap offers the API through the COM. This offering is also known as exposing the API. Using ArcMap to write your own spatial query application achieves one of the main COM objectives: software reuse. You reuse ArcMap's query routines instead of writing your own.

A software component is a self-contained and reusable binary code that is independent of its programming language and its location. Software components are stored in either dynamic link libraries (DLLs) or executable files. The file that contains the components is also referred to as a server file. The location independency is also known as location transparency, and refers to where the component is stored and in what form (for instance, as a DLL or executable).

There are variations of the COM, which is known by different names. DCOM and COM+ are variations of the COM. The D in DCOM stands for "distributed." DCOM makes COM-based software available across a network. COM+ is the second-generation COM. It has the DCOM along with application server technology. ActiveX and OLE are other names used for the COM.

Microsoft recently changed the original meaning of ActiveX to be a brand name for all of Microsoft's COM-based technologies. Microsoft has also changed OLE's meaning to be the COM-based technologies that implement compound documents. The compound documents can store documents from different applications. For example, you can store an Excel spreadsheet inside a Word document.

Programming Language Independence

Programming language independence means that you can use any programming language to use the components, regardless of the programming language used for the components. COM specifications identify two factors for programming language independence. One is defining the software component independently of the programming language. The other is to adhere to binary standards that specify how a component can be accessed.

Type libraries are the way software components are defined independently of the programming language. A type library is a binary file that describes your server in detail. It has the name of the classes, the class interfaces, interface methods and properties, parameters for the methods, and the globally unique identifiers (GUID) for each class or interface. An interface is a group of public methods defined in a class. Interfaces, methods, and properties are described in Chapter 6.

A GUID (rhymes with squid) is a 16-byte integer that uniquely identifies a class or interface. It is important to have a different identifier for each class and interface so that components do not conflict. Defining a GUID as a 16-byte integer provides for practically an infinite number of possible combinations. The following is an example of a GUID value.

```
{5DEB1DB8-C2A9-11D1-B9A2-080009EE4E51}
```

The GUID value is placed within brackets. The preceding GUID value is for the standard toolbar in ArcMap. The ArcObjects developer help file provides a complete list of GUID values for ArcMap and ArcCatalog components. There is a standard algorithm for creating a GUID. Many of the programming tools, such as Visual Studio, have utilities that can create a GUID. Current date and time, a machine identifier, and a counter are used in building the GUID value.

You have to take certain actions to make a type library part of your application if you are using Visual Basic or C++ to build an ArcObjects application. In VBA, the type libraries are included as part of your project and there is no action required. VBA's IntelliSense helps you complete a statement by showing you the list of possible methods or properties, or by listing the set of parameters. IntelliSense gets its information from the type libraries.

In addition to having a type library, a COM server must provide the ability for other applications to call its components independently of the programming language. This requires a binary-level calling standard. The standard specifies that COM interfaces must be implemented as an array of function pointers. Each method in the interface has a function pointer in the array.

Location Transparency

Location transparency means that you can use all COM components in the same manner, regardless of the server type that is DLL or executable, or of server location. When the COM server is in the form of a DLL, your application can easily access its interfaces because DLL runs in the same process. However, if the COM server is in form of an executable file, it cannot run in the same process as your application, and therefore your application cannot access its interfaces.

Marshalling is the process that creates location transparency. Marshalling allows an application to interact with the interfaces of a COM server that is not in the same process. The marshalling process creates a proxy for the COM server. The application then goes through the proxy to interact with the COM server.

As a VBA programmer, you are not creating COM servers. However, understanding the basics of COM helps you in developing ArcObjects applications. Chapter 6 focuses on the COM interfaces. Developing via ArcObjects is synonymous with developing via interfaces because ArcObjects exposes only its interfaces.

Component Object Model Interfaces

AN INTERFACE IS A GROUP OF RELATED METHODS that interact with an object. A method, also known as a function, carries out a specific task. For example, the *Map* object in ArcMap has over twenty interfaces. These interfaces group related methods. For instance, the *IGraphicsContainer* interface groups the methods that control the graphics container of the *Map* object. *DeleteAllElements* is one of the methods; it deletes all graphic elements of a map.

You need to have a good understanding of interfaces because these represent the means of interacting with ArcObjects. This chapter describes the interface concept. Chapter 7 shows you how to use interfaces when writing ArcObjects applications.

Classes and Objects

If you have been developing applications with Visual Basic or C++, you have been more interested in classes than interfaces. In fact, you may not have had any reason to deal with interfaces up to this point. Before delving into the interface concept, you should have an understanding of the classes and objects. The reason is that interfaces organize methods that are exposed by objects, and objects are the instantiated forms of classes.

Classes and objects came into the software developer's lexicon with object-oriented programming (OOP) technology. Classes provide the blueprint for creating objects, and objects are the things you work with to build an application. An object in an application is a piece of code that maintains some property values and can operate on those values through its methods.

A class defines the properties and methods of an object. Not every class can instantiate an object. Often classes are used so that other classes can be defined. An abstract class is the type of class that cannot instantiate an object but is used to define other classes. Certain classes have objects but cannot instantiate them. In such cases, the object has to be accessed through another object. A class that can create an object is referred to as a co-class. For example, the *LineElement* co-class can create objects that are graphic lines on a map.

Interfaces

An object in ArcObjects may expose several methods. The methods are used like a function to operate on the object or carry out a specific task. An interface is a group of related methods. When you develop ArcObjects applications, you do everything through interfaces. In the following code, the name of the spatial reference for a geo-dataset is displayed in a message box. This example shows how interfaces are used to access a layer's geo-dataset.

```
Dim pMxDocument As IMxDocument
Dim pMap As IMap
Dim pGeoDataset As IGeoDataset
Dim pSpatialReference As ISpatialReference
Set pMxDocument = Application.Document
Set pMap = pMxDocument.FocusMap
Set pGeoDataset = pMap.Layer(0)
Set pSpatialReference = pGeoDataset.SpatialReference
MsgBox pSpatialReference.Name
```

The preceding code segment starts by declaring four variables that reference interfaces. The declaration statements begin with the *Dim* keyword and end with the referenced interface. For example, *Dim pMap As IMap* declares the *pMap* variable as a type that references the *IMap* interface. The variable actually references the *IMap* interface of the activated map once the *Set pMap = pMxDocument.FocusMap* statement has executed.

The *pMxDocument.FocusMap* expression returns a reference to the *IMap* interface of the activated map in the ArcMap document. Once *pMap* is pointing to the *IMap* interface, you have access to the *Map* object. You never deal directly with the *Map* object; instead, you go through its interfaces (such as *Imap*).

The *IMap* interface offers a property named *Layer* that returns a reference to an interface of a layer. The *Set pGeoDataset = pMap.Layer(0)* statement sets the *pGeoDataset* to reference the *IGeoDataset* interface of the first *Layer* object in the activated map. The *pMap.Layer(0)* expression actually returns a reference to the *ILayer* interface. However, the statement assigns the reference to a variable of type *IGeoDataset*. This is possible because both *ILayer* and *IGeoDataset* are interfaces of the same *Layer* object.

You can access a new interface through a referenced interface of the same object. In VBA, this action is done through a hidden method known as *QueryInterface*. To access an interface through a referenced interface, you declare the new interface and then set it equal to the referenced interface. All COM objects, without exception, offer the *QueryInterface* method. This method is offered through the *IUnknown* interface that all COM objects implement.

In the preceding code example, the reason we are interested in referencing the *IGeoDataset* interface is that it has the property *SpatialReference*, whose name we want to display. As you can see, the example never accesses the *Map* or *Layer* object directly. Instead, it uses the objects' interfaces to interact with the objects.

Inbound and Outbound Interfaces

An interface can be inbound or outbound. Your VBA application can call the methods of an inbound interface to perform tasks. In ArcObjects, most interfaces are inbound. For example, your application can at any time use the *DeleteLayer* method of the *IMap* interface to remove a layer from the *Map* object. The *IMap* is an inbound interface.

Methods in an outbound interface work in a manner that is the reverse of the inbound interface process. These methods call on your application to perform tasks. Generally, outbound interfaces group methods that handle event processing. For example, the *MxDocument* that represents an ArcMap document has an outbound interface named *IDocumentEventsDisp*. This interface groups methods that listen for specific events related to the map document. For instance, the *OpenDocument* method is executed when a map document is opened. The following steps take you through the process of creating a procedure in VBA that will display a message each time your document is opened.

1 Start ArcMap and save it as a new map document.

2 Open the VBA editor by pressing the Alt + F11 keys.

3 In the VBA editor's Project Explorer window, expand the Project and ArcMap Objects nodes so that you can see the *ThisDocument* object. If Project Explorer is not displayed, select View | Project Explorer to open it.

4 Double click on the *ThisDocument* object to open the Code window.

5 In the Code window, select *MxDocument* from the Objects drop-down list. Then select *OpenDocument* from the Procedure drop-down list. This action places the following function stub in the Code window.

```
Private Function MxDocument_OpenDocument() As Boolean
End Function
```

6 Add the *MsgBox* statement to the function stub so that the function appears as follows.

```
Private Function MxDocument_OpenDocument() As Boolean
MsgBox "Hello world."
End Function
```

This function is executed each time you open this document. Test the function by saving the map document and reopening it.

In this and the previous chapters you learned about COM basics. You also learned that ArcObjects applications are developed by accessing and working with COM interfaces. In Chapter 7 you will learn how to apply your understanding of COM to the creation of ArcObjects applications.

CHAPTER 7

Developing COM Applications

ARCOBJECTS IS BASED ON MICROSOFT COM technology. You can develop ArcObjects applications to customize ArcGIS using any programming language that supports COM. This book shows you how to develop applications using VBA, which is integrated with ArcGIS. However, you can also develop ArcGIS applications with Visual Basic or C + + .

This chapter covers topics related to developing ArcObjects applications and macros with COM. At the end of the chapter, you are presented with a set of programming guidelines. You can improve readability of your code and make it easier to maintain by adhering to these guidelines.

Where to Start

You develop ArcGIS applications and macros by manipulating ArcMap or ArcCatalog objects through their interfaces. In VBA applications, generally, you start with the *Application* object and go through the object model until you reach the desired object. The *Application* object is available to all VBA procedures. You do not need to declare and reference it. In the following example, the *Application* object is used to access the ArcMap document.

```
Dim pMxDocument As IMxDocument
Set pMxDocument = Application.Document
```

In the preceding example, the reference to the ArcMap document is declared and then set. After you have set the reference (second line), you can use the *pMxDocument* variable in your application. Using the reference to the ArcMap document you can then reference the activated map using the *FocusMap* property of the *IMxDocument* interface. From the activated map, you can reference the layers, and from the layers you can reach the features.

The *Application* object offers access to the *IApplication* interface. If you need a different interface for the application, you can access it by querying the *IApplication* interface, as shown in the following example. The following code segment queries the *IApplication* interface to reference the *IWindowPosition* interface. The *IWindowPosition* is one of the application's interfaces. Next, the *Move* method of the *IWindowPosition* is applied to set the position and size of the application's window.

```
Dim pWindowPosition As IWindowPosition
Set pWindowPosition = Application
pWindowPosition.Move 1, 1, 500, 500
```

Checking the Interface Type

Often your application may reference an interface that could be one of several types. For example, you can reference the first layer in a map, but the layer could be a feature or image layer. You can use the *TypeOf* keyword to determine your variable's interface type. The *TypeOf* operator returns *True* if the variable can query the specified interface. The following code segment shows you how to determine if a layer is a feature layer. The code segment retrieves the reference for the first layer of the activated map. It then checks to see if the layer is a feature layer.

```
Dim pMxDocument As IMxDocument
Dim pLayer As ILayer
Set pMxDocument = Application.Document
Set pLayer = pMxDocument.FocusMap.Layer(0)
If TypeOf pLayer Is IFeatureLayer Then
   ' Your code goes here
End If
```

The *TypeOf* keyword is used with the *Is* operator. You can use the *Is* operator to compare two interfaces, or to check if a reference to an interface is valid. A common statement in ArcObject's applications is to validate the reference to an interface. For example, you may try to reference a feature in a layer but because the feature does not exist the reference is invalid. You can test the validity using the *Is* operator with the *Nothing* keyword. In the following example, the validity of a reference to a feature's geometry is tested.

```
If pGeometry Is Nothing Then
  ' Your code for invalid interface
End If
```

The *Is Nothing* expression returns *True* when the reference is invalid. The *Is* operator can also be used to compare two variables that reference interfaces. It would return *True* if both variables are referencing the same interface.

Client-side Storage

A few of the interface methods that return a reference to a pointer require a valid pointer in their parameter list. This is known as client-side storage, in that the client is reserving the memory needed for the variable. In the following example, *QueryEnvelope* returns the extent of *pPolygon* in the *pEnvelope* variable.

```
Dim pEnvelope as IEnvelope
Set pEnvelope = New Envelope
pPolygon.QueryEnvelope pEnvelope
```

In the preceding example, *pEnvelope* is constructed on the second line to reference a new and valid *Envelope* object. Next, the *QueryEnvelope* method populates it.

Using Enumerators

Enumerators facilitate iterating through a collection in COM. Collections are objects that contain more than one object of the same type. For example, the *Layers* property of the *IMap* interface returns a reference to *IEnumLayer*. You can iterate through the layers of a map with *IEnumLayer*. The enumerators generally have at least two methods: *Reset* and *Next*. The following code segment shows you how the *IEnumLayer* interface is accessed and used.

```
Dim pMxDocument As IMxDocument
Dim pMap As IMap
Dim pEnumLayer As IEnumLayer
Dim pLayer As ILayer
Set pMxDocument = Application.Document
Set pMap = pMxDocument.FocusMap
Set pEnumLayer = pMap.Layers
pEnumLayer.Reset
Set pLayer = pEnumLayer.Next
Do While Not pLayer Is Nothing
  MsgBox pLayer.Name
   Set pLayer = pEnumLayer.Next
Loop
```

In the preceding example, *pEnumLayer* references the *IEnumLayer* interface. The *Layers* property of the activated map returns the reference to *IEnumLayer*. Before iterating through an enumeration, you should first apply the *Reset* method. This method sets the enumeration sequence to the beginning. The reason for this action is that depending on how the enumerator is created its sequence may or may not be at the beginning.

The *Next* method returns a reference to the current interface and increments the sequence to the next member. After you have applied the *Reset* method, the *Next* method returns the first member of the enumeration. Applying the *Next* method again returns the second member. The *Next* method returns *Nothing* when there are no more members.

The loop structure in the preceding example is a typical way of iterating through members of an enumeration. Before the loop, you apply the *Reset* method and get the first member. Inside the loop you apply the *Next* method so that the loop iterates through the enumeration. If you leave out the *Set pLayer = pEnumLayer.Next* statement, you will have an infinite loop because it continues until the returned object is *Nothing*.

You can also directly access a member of a collection using the member's index number. In this manner you do not use an enumerator. Instead, you use properties that give you access to the collection's members. For example, you can reference the first layer in a map with an enumerator by using the *Layer* property, as shown in the following code segment.

```
Dim pMxDocument As IMxDocument
Dim pLayer As ILayer
Set pMxDocument = Application.Document
Set pLayer = pMxDocument.FocusMap.Layer(0)
```

The *Layer* property accepts a position index as its parameter and returns the *ILayer* in that position. All collections in ArcObjects start with the position index of zero. Your application will produce a run-time error if you ask for a position index that does not exist. Therefore, you should always be aware of the number of members in a collection. For example, the *LayerCount* property of the *IMap* interface returns the number of layers in the layer collection.

Programming Guidelines

Using programming guidelines or standards makes your applications or macros easier to understand and maintain. The guidelines presented here are suggestions, and you can change or add new standards. The important factor is to be consistent. The applications and macros developed in this book follow these guidelines where applicable.

Procedure Name

Each VBA macro or procedure has a procedure that is executed when you run your application. This procedure's name should match your application's name. An application may also have other procedures that perform specific tasks. For example, a procedure may clear a layer's selection. The name for those procedures should start with a verb describing the operation and end with one or more words describing the operation's subject, such as *ClearLayerSelection*.

Functions return a value or object. Therefore, start their names with *Return*. For example, *ReturnFirstSelectedFeature* indicates that the function returns the first selected feature. When a function returns a Boolean value as a test, start the function name with the word *Is*. For example, the *IsLayerLoaded* function returns *True* if a specific layer is loaded. Otherwise, it returns *False*.

Keep in mind that VBA has a specific naming standard for handling procedures that respond to events. For example, *UIToolPermit_MouseUp* is executed when you release the mouse button. You should not change the procedure names established by VBA.

Header

The header is the block of comment lines at the beginning of each procedure. Its purpose is to provide a quick description of the procedure and any additional information helpful in using or maintaining the code. If you are writing a small macro, a short and brief header is sufficient. For larger applications, however, consider a more detailed header.

Comment Line

Comments within your application are crucial to understanding and maintaining the code. Do not repeat what the code shows; instead, explain the reason for the action. Use a blank line to attract the reader's attention to the comment. If you are placing a comment in an indented block, indent the comment as well. The following is an example of a bad comment, in that it only reiterates the action of the *If* statement and does not explain the process.

```
For lngIndex = 0 To pMap.LayerCount - 1
' Check the name of the layer
   If pMap.Layer(lngIndex).Name = strLayerName Then
     Set GetLayer = pMap.Layer(lngIndex)
     Exit For
   End If
Next lngIndex
```

A better comment is used in the following code segment.

```
For lngIndex = 0 To pMap.LayerCount - 1
  ' Search through layers for the given layer name
  If pMap.Layer(lngIndex).Name = strLayerName Then
    Set GetLayer = pMap.Layer(lngIndex)
    Exit For
  End If
Next lngIndex
```

Variable Name

Start a variable name with one or more lowercase characters that represent the variable's type. Use meaningful names that describe the content of the variable. If you use abbreviations, be consistent. The following are examples of variable name use.

- Variable name *pStateFeatureLayer* indicates it has a pointer to the *IFeatureLayer* interface of the *State* layer.

- Variable name *sLayerName* or *strLayerName* indicates that it contains a layer's name in text string format.

- Variable name *frmUserInput* could be used to reference a user form.

When you declare variables, you must place the declaration prior to the statement that uses the variable. The declaration can appear at the beginning of the procedure or in the middle of the procedure. It is recommended that you place all declarations at the beginning of the procedure. This way, you search only one area of the code for a declaration. However, in some procedures of this book the declaration is placed near the statement that uses the variable so that blocks of code can be explained without having to jump back and forth among parts of the procedure.

Alignment

Indent the statements inside conditional or loop structures. If you break a statement into more than one line, align the following line with the first line. The following code segment is an example of indentation.

```
If pLayer Is Nothing Then
  MsgBox "Unable to locate " & strDataPath & _
  strTractFileName & " shape file."
  Exit Sub
End If
```

CHAPTER **8**

ArcObjects Basics

ARCOBJECTS IS THE SOFTWARE DEVELOPMENT PLATFORM for the ArcGIS family of applications. It is the collection of tools, libraries, and environments you use to customize, extend, or build extensions for ESRI software products. ArcObjects is based on Microsoft's Component Object Model (COM) technology. Therefore, you can use any COM-compliant development language, such as Visual Basic or C++, to build ArcGIS applications.

This book covers using the Visual Basic language in the Visual Basic for Applications (VBA) environment to develop ArcMap macros and applications. The VBA environment was described at the beginning of this book. In this chapter, you are introduced to various object libraries and tools available through ArcObjects.

ESRI Object Libraries

An object library is a file that contains information on a collection of objects available to programs. These objects are units of code and data that perform distinct tasks. You include these libraries in your application so that you can access their objects. ArcObjects has organized its objects in several different libraries. The primary libraries are described in the following.

- The Core Object Library should be included in all of your applications. This library incorporates objects related to many areas, including Application Framework, ArcCatalog, ArcMap, Geocoding,

Geodatabase, and Geometry. The Core Object Library is automatically included in your VBA environment. This library is stored in the *esriCore.olb* file.

- The ArcMap Object Library contains the Application co-class for ArcMap. It must be included in all of your applications for ArcMap. This library loads automatically when working with VBA in ArcMap and is stored in the *esriMx.olb* file.

- The ArcCatalog Object Library is stored in the *esriGx.olb* file. It contains the Application co-class for ArcCatalog. This library loads automatically when working with VBA in ArcCatalog.

- Use the ESRI ArcObjects Controls 8.1 when developing mapping applications that are independent of ArcMap. This library is stored in the *afcontrols.ocx* file.

- The UIControls Library contains objects that represent pushbuttons, combo or edit boxes, and tools on a custom toolbar or dialog box. These objects can be used within the VBA environment only. The UIControls Library is stored in the *UIControls.dll* file.

You can verify that the required libraries are included in your applications by reviewing the libraries referenced by your VBA project. In the VBA editor, select Tools | References to display the References dialog box, shown in figure 8-1.

In figure 8-1, the libraries that have a check box are included in your application. To add a new library, find it on the list and place a check mark next to it. You should not include object libraries you do not use because this adds to the number of object references VBA must resolve, causing it to slow down. VBA automatically loads certain object libraries when you start the VBA editor. Do not remove libraries automatically loaded.

The IntelliSense functionality of VBA is the list box of valid options that appears while you are writing your VBA code. IntelliSense reads the object libraries included in your application to determine the valid options. Figure 8-2 shows an example of how IntelliSense works.

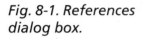
Fig. 8-1. References dialog box.

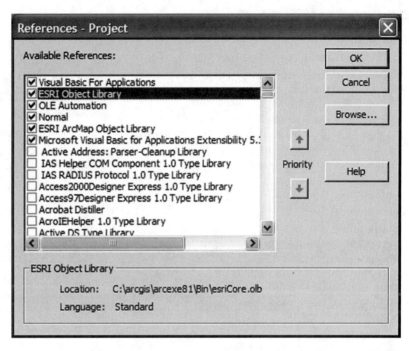

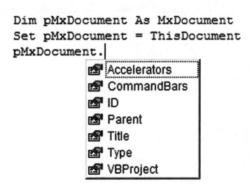

Fig. 8-2. Example of VBA's IntelliSense.

You can see the content of an object library by using VBA's or ESRI's object browser utilities. You can access VBA's object browser by selecting View | Object Browser. The next section explains how to start and use ESRI's object browser.

ESRI's Object Browser

ESRI provides an object browser utility with ArcGIS named EOBrowser (ESRI Object Browser). Although you can use the VBA's object browser, ESRI's utility is customized to work well with ESRI's object libraries.

Start EOBrowser by running the *EOBrowser.exe* program from the *\ArcObjects Developer Kit\Utilities* subdirectory of your ArcGIS installation. It is recommended that you create a shortcut for this program so that you do not have to use Windows Explorer to find the file each time you need it. EOBrowser is shown in figure 8-3.

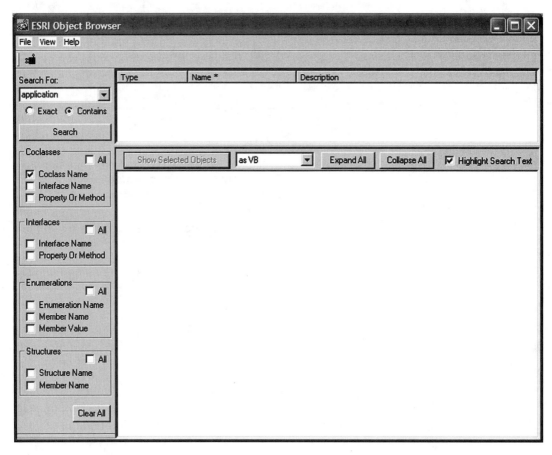

Fig. 8-3. EOBrowser.

You must first load a library before you can browse its objects. Select File | Object Library References to open the Object Library References dialog box, shown in figure 8-4. In that dialog box the libraries that are loaded are listed with a check mark next to them. In figure 8-4, the Core Object Library is loaded.

Fig. 8-4.
Object
Library
References
dialog box.

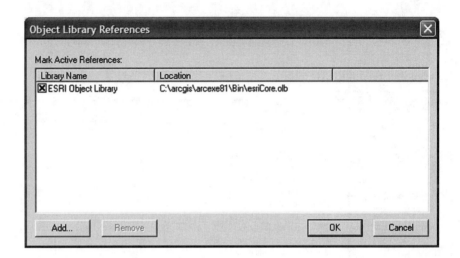

The following steps take you through the process of loading an object library into the ESRI Object Browser.

1 Select File | Object Library References to display the Object Library References dialog box.

2 Click on the Add button to display the Select Type Library dialog box.

3 Use the drop-down list from the registered type libraries to find the desired library. This list contains all registered object libraries, in addition to the type libraries. Click on the OK button.

4 Click on the OK button in the Object Library References dialog box.

  TIP: *ESRI's Object Browser is not limited to ESRI libraries. You can use it to browse any object library.*

A library may contain hundreds of objects. To view an object, you can search for it using the controls on the left-hand side of the EOBrowser window. When you place a check mark next to one of the All options, all corresponding objects are displayed, regardless of the search criterion. Figure 8-5 shows a search for the measurement unit enumeration using *inches* as the search criterion.

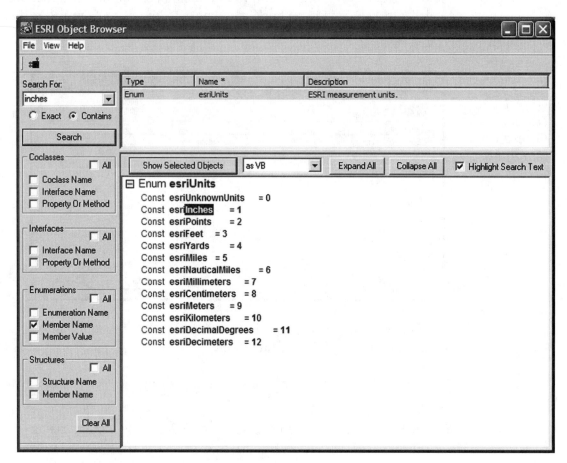

Fig. 8-5. Use of the Search for Measurement Unit Enumeration function.

The EOBrowser places the search results in the upper list box. Select one or more objects from the upper pane and click on the Show Selected Objects button to display the object's detail in the lower pane. Figure 8-6 shows another search example for determining how you can get the full extent.

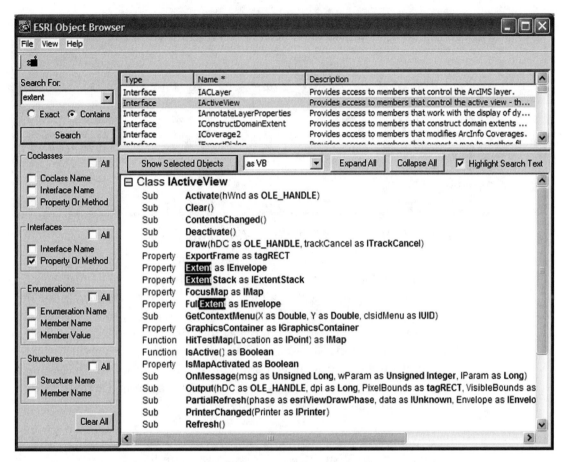

Fig. 8-6. Use of the Search for Extent function.

In figure 8-6, both methods and properties are shown for the *IActive-View* class. The methods are identified by the keyword *Sub* or *Function*.

ArcObjects Developer Kit

The ArcObjects Developer Kit consists of help files, sample code, diagrams, and utilities stored in the *ArcObjects Developer Kit* directory of the ArcGIS installation. Figure 8-7 shows you the list of subfolders in the Developer Kit that organize the kit's resources. If you do not have these folders, you need to perform a full or custom ArcGIS installation.

Fig. 8-7. ArcObjects Developer Kit folders.

The *Help* folder contains the ArcObjects developer help file. You can access this help by selecting Start | ArcGIS | ArcObjects Developer Help or by opening the *AODEV.chm* file in this directory. The *Kits* folder contains the source code in Visual Basic and Visual C + + for several of the ArcMap commands. These source codes can be very useful in understanding and creating new commands. The *Object Model Diagrams* folder contains all ArcObjects object model diagrams. Chapter 9 shows you how to read and use these diagrams.

The *Samples* folder contains many ArcGIS applications written in Visual Basic or Visual C + +. You should familiarize yourself with the files in this folder. Often the application or macro you need to write is partially or fully written and available as a sample. Furthermore, studying the samples teaches you how to develop ArcObjects applications.

The *Utilities* directory contains several useful development tools (such as the ESRI Object Browser discussed previously) and tools for fixing corrupt registry or ArcMap files. The ArcObjects Developer Kit can help you learn and develop ArcObjects applications.

CHAPTER 9

ArcObjects Elements

CHAPTER 8 INTRODUCED YOU TO THE ARCOBJECTS libraries. You have to use one or more of these libraries in every ArcGIS application or macro you write. The libraries contain the objects you need to access and manipulate to carry out a task in your program. Often you need to start with an available object to be able to access another.

Information about each object and the relationships among the objects are graphically presented in the ArcObjects object model diagram. Understanding the object model diagram helps you write better applications. This chapter teaches you how to read the ArcObjects object model diagram. It also shows you how to employ a methodology associated with the ArcObjects object model for writing VBA applications.

Reading the ArcObjects Object Model Diagram

ESRI provides the ArcObjects object model diagram in the ArcObjects developer help and various other publications. The diagramming notation is based on the Unified Modeling Language (UML), which is a standard diagramming language for object-oriented design.

The entire object model is too voluminous to present in this book. However, segments of it are shown as applicable. You can see the entire model by opening the *AllOMDS.PDF* file in the *Object Model Diagrams* directory of the ArcObjects Developer Kit. Learning how to read and use the object model diagram helps you in writing ArcGIS applications. You could avoid the object model when customizing the interface or writing small macros in VBA, but for complete and complicated applications you need to use the object model.

The object model diagram displays classes and objects, relationships among classes, class interfaces, and interface details. A diagram may be shown with some of its details removed for clarity. It would make a very busy and unusable diagram if all details were shown together. This book shows the object model diagram at three levels of detail. A diagram might show only applicable classes and their relationships (as in figure 9-3), a single class and its interfaces (as in figure 9-6), or the details of a single interface (as in figure 9-7).

Objects

Objects are derived from classes. The ArcObjects object model diagram shows the classes that can create or instantiate objects. Showing the objects in a diagram is therefore an example of a class diagram as a snapshot of the executed application. In other words, it represents what the system constitutes at any given moment in its development. This book shows only the classes in the model, which is almost identical to a diagram of objects. In an object-oriented environment, the classes are considered special types of objects, and thus the name *object model diagram*.

Classes

Descriptions of the three types of classes follow.

• An abstract class cannot create objects while its descendents have objects. The purpose of an abstract class is to model common properties and methods of several classes into one. *Command* is an example of an abstract class. You cannot create an object from the *Command* abstract class. However, you can have objects from *Command*'s descendents, such as a tool or button on a toolbar.

- A co-class can create new objects. For example, the *FeatureLayer* co-class can create feature layer objects you can display and manipulate.

- A class cannot create a new object directly. However, a class's object can be derived from other objects. For example, the *CommandBars* class represents the collection of toolbars in a document. You cannot create a new object of type *CommandBars*, but once you have the document object you can access the *CommandBars* object through the *CommandBars* property of the document object.

Figure 9-1 shows how each class type is displayed in the object model diagram.

Fig. 9-1. Class types.

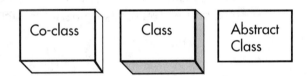

The following code segment is an example of how a co-class is created in VBA.

```
Dim pFeatureLayer As IFeatureLayer
Set pFeatureLayer = New FeatureLayer
```

Classes cannot be created directly; they are derived from other objects. The following code segment is an example of how an object of class *CommandBar* is accessed.

```
Dim pMainMenuBar As ICommandBar
Set pMainMenuBar = Application.Document. _
CommandBars.Find(ArcID.MainMenu)
```

Relationships

Classes exhibit relationships among one another, also known as association. There are several distinct types of associations, described in the following.

- Inheritance is presented with a triangle, as shown in figure 9-2. The descendent classes share the same properties and methods as their ancestors. Figure 9-2 shows how the abstract class *Layer* specializes into *FeatureLayer* and *CadLayer* co-classes. The two co-classes inherit properties and methods from the *Layer* abstract class.

- The aggregation relationship is a part/whole association presented with a diamond, as shown in figure 9-3. The aggregation indicates that several classes combined make a new class. In figure 9-3, you can see that an ArcMap document co-class is a combination of table of contents, map, and page layout. The integration association is called composition when objects from the whole class control the lifetime of part classes.

- The instantiation relationship shows how one class's property or method can create objects of another class. Figure 9-4 shows the instantiation relationship between a co-class and class.

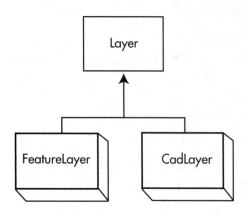

Fig. 9-2. Inheritance relationship.

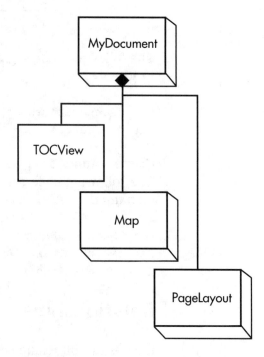

Fig. 9-3. Aggregation relationship.

Fig. 9-4. Instantiation relationship.

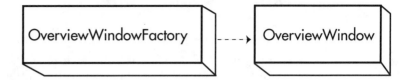

Interfaces

Developing ArcObjects applications involves using interfaces. You manipulate and communicate with objects through their interfaces. There are two types of interfaces: inbound and outbound. Figure 9-5 shows how an interface is displayed in the model.

Fig. 9-5. Class interface.

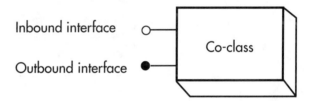

Most interfaces in ArcObjects are inbound. Your application calls the functions of an inbound interface. The outbound interfaces have functions that call your applications. The outbound interfaces are used to handle event-driven functions. Figure 9-6 is an example of how interfaces are displayed for the *Page* co-class.

Fig. 9-6. Page co-class interface.

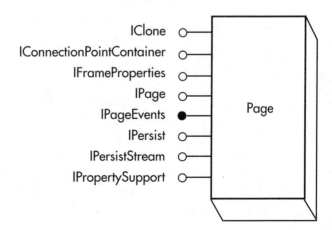

Methods and Properties

Each interface offers methods and properties that facilitate access and manipulation of objects. Methods carry out an action, and may require some parameters and may return some values. Properties are attributes of an object. You can set some properties, but others are read-only. Methods are shown with arrows, and properties are shown using squares, as shown in figure 9-7.

Fig. 9-7. Methods and properties.

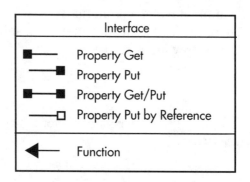

The *get* property is equivalent to a read-only property, whereas the *put* property is write-only. A *get/put* property allows you to read and write the value of the property. For example, the *FeatureLayer* co-class has an interface called *IFeatureLayer*. This interface has inherited a *get/put* property named *Visible* from the *Layer* abstract class. The following VBA macro shows you how to read or write a property by first examining the value of the visible property and then setting its value.

CODE VBA09-1

```
Public Sub MakeLayerVisible()
 Dim pMxDocument As IMxDocument
 Dim pMap As IMap
 Dim pFeatureLayer As IFeatureLayer
 Dim pActiveView As IActiveView
 Dim pContentsView As IContentsView

 ' Access the first feature layer on the map.
 Set pMxDocument = ThisDocument
 Set pMap = pMxDocument.FocusMap
 Set pFeatureLayer = pMap.Layer(0)
```

```
' If the feature layer is not visible,
' then make it visible.
If Not pFeatureLayer.Visible Then
 pFeatureLayer.Visible = True
End If

' Refresh the map.
Set pActiveView = pMap
pActiveView.Refresh

' Refresh the table of contents.
Set pContentsView = pMxDocument. _
CurrentContentsView
pContentsView.Refresh pFeatureLayer
End Sub
```

When a property is designated as *put* by reference, you need to use the *Set* keyword to assign a value to it. The following code segment is an example of how to assign a value by reference. In this example, a feature layer's class is set by reference using the *FeatureClass* property.

```
Set pFeatureLayer = New FeatureLayer
Set pFeatureLayer.FeatureClass = pFeatureClass
```

Methodology for Writing ArcObjects Applications

Similar to computer applications in other fields, GIS applications must be developed according to a structured methodology. There are a variety of methodologies that encompass the entire software development lifecycle. They include stages for requirements analysis, design, construction, and testing. Covering the entire software development lifecycle is beyond the scope of this book. However, this section offers a structured approach to constructing a VBA application or macro for ArcGIS.

For many people, methodology implies extensive overhead, extraneous tasks, and the stifling of creativity. In reality, however, adopting a methodology simply means using certain techniques and executing certain tasks that will produce the best results. If you plan to write short VBA macros to automate a few tasks, you may apply only parts of the methodology that are applicable. If you plan to develop complete applications, following a structured methodology dramatically increases the probability of a successful outcome.

The methodology presented here (developed by the author) might be referred to as "break down and assemble." It follows the divide-and-conquer concept. Similar to many complicated operations, breaking the whole into smaller and simpler pieces helps in completing the task. The process breaks down the application into manageable parts that you construct and then assemble to build the application. Figure 9-8 shows a graphical representation of this methodology.

The following sections describe each stage of the methodology using an example of an ArcMap application named *GoCounty*. The *GoCounty* application helps a user navigate to a U.S. county and then exports the displayed map to a PDF file. The application loads the *County* layer if this layer is not available.

Stage 1: Definition

You should not even think how you would construct the application until you have a full understanding of the application. Whether you are developing for your own use or for someone else, you should have the application's definition on paper. Committing the definition to paper organizes your thoughts and improves clarity. You should record the definition in plain English in the form of an outline. The definition for the example *GoCounty* application follows.

- Start the process when the user selects the application's command.

- First make sure that the *County* layer is available. If it is not, load it.

- Display a user form with a list of selectable counties.

- The application pans and zooms to the selected county.

- The application exports the map to a PDF file.

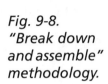

Fig. 9-8.
"Break down
and assemble"
methodology.

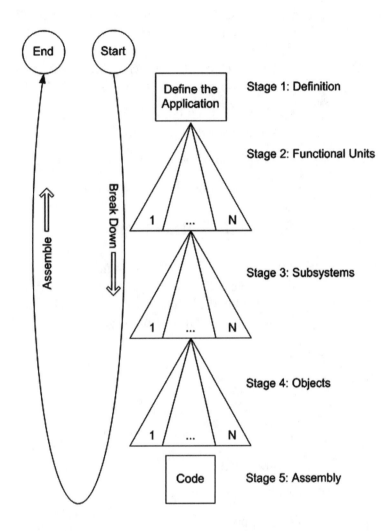

Stage 2: Functional Units

At this stage, you break the application into functional units. Each functional unit executes part of the application. The outline from the application definition is a good guide for identifying the functional units. Define the units as independent mini-applications. When documenting the units, try to use ArcGIS terminology, which helps in the next stage. In addition, try to list the units in the order in which they will be processed within the application. The functional units for the *GoCounty* application are as follows.

- *Unit 1:* Check the ArcMap document for availability of the *County* layer and load it from the personal Geodatabase if it is not found.

- *Unit 2:* Extract county names from the *County* layer and populate the user form with the names.

- *Unit 3:* Find the county feature and set the map extent to it.

- *Unit 4:* Export the map to a PDF file.

Stage 3: Subsystems

Take each functional unit identified in stage 2 and break it down into the subsystems required to support the unit. Subsystems are defined as sets of related objects within the ArcObjects object model. The following are the available subsystems.

- ArcCatalog
- ArcMap
- ArcMap Editing
- Display
- Framework
- Geodatabase
- Geometry
- Network
- Output
- Raster
- Spatial Reference

Open the *ArcGIS Object Model.pfd* file in the *Objects Model Diagram* folder of the ArcObjects Developer Kit to see the objects associated with each subsystem. For the *GoCounty* example, you need to take each functional unit and break it down into its subsystems. To carry out unit

3, you need to query the attribute data in the personal geodatabase. You then need the geometry of the selected county. Finally, you need to change the extent of the map so that the selected county is displayed. Accordingly, the following subsystems are needed to support unit 3.

- *Subsystem 1:* Geodatabase

- *Subsystem 2:* Geometry

- *Subsystem 3:* ArcMap

Stage 4: Objects

This stage breaks down the result of the previous stage. For each subsystem in stage 3 you need to identify the objects you need to carry out a given task. If you are not sure about the required objects, review the object model and read about candidate objects in the help system. Once you have developed a few applications, this stage becomes easier. Breaking down the first subsystem of functional unit 3 yields the following objects.

- *QueryFilter co-class:* To search based on attribute values.

- *FeatureCursor class:* To hold the result of query.

- *Feature co-class:* To access the geometry of the selected county.

You then need to identify the input and output of your subsystem (in this case, as follows). The input, if any, shows the objects your subsystem needs to complete its task. Because by definition functional units are independent of one another, the input to the subsystem must come from the output of another subsystem in the same unit. The output, if any, is the result of your subsystem. In the *GoCounty* example, the input is the *County* layer so that you can perform a query on its feature class. Output is the feature for the selected county.

- *Input:* County layer

- *Output:* Selected county feature

Stage 5: Assemble

At this stage you actually write the ArcObjects code. The code generally consists of first establishing the object by either creating it or deriving it from another object and then accessing the object's parameters or performing an operation through its methods. You assume that some subsystem in one of the functional units will provide the input identified in stage 4. In the same manner, this subsystem's output will become some other subsystem's input.

The outcome of the assembly stage for each subsystem is a VBA procedure. Once you have completed the procedures for all subsystems of a functional unit, you write the code for the functional unit to assemble the procedures of the subsystems. The result is a VBA procedure that codifies the functional unit. This procedure generally consists of calling the subsystem procedures and managing their input and output.

Once the procedures for all functional units are completed, you write the VBA procedure for the application that calls each functional unit to perform its tasks. Figure 9-9 depicts how the application code assembles and calls its procedures.

Fig. 9-9. Procedure assembly.

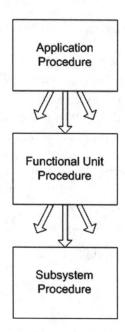

The following procedure is for the *GoCounty*'s subsystem 1 (Geodatabase) of the third functional unit.

CODE
VBA09-2

```
Private Function GetCountyFeature( _
pFeatureLayer As IFeatureLayer, _
strCountyName As String) As IFeature

   ' Query the feature class.
   Dim pFeatureClass As IFeatureClass
   Dim pQueryFilter As IQueryFilter
   Dim pFeatureCursor As IFeatureCursor

   Set pFeatureClass = pFeatureLayer.FeatureClass
   Set pQueryFilter = New QueryFilter
   pQueryFilter.WhereClause = "NAME = '" & _
     strCountyName & "'"
   Set pFeatureCursor = pFeatureClass.Search _
     (pQueryFilter, False)

   ' Get the feature.
   Dim pFeature As IFeature

   Set pFeature = pFeatureCursor.NextFeature
   If pFeature Is Nothing Then
     Set GetCountyFeature = Nothing
   Else
       Set GetCountyFeature = pFeature
   End If
End Function
```

The preceding VBA function accepts as input the *County* layer and the name that user has selected. The function finds the county and returns its feature. The procedure for the functional unit could call the subsystem with the following code segment.

```
Dim pFeature As IFeature
Set pFeature = GetCountyFeature(pFeatureLayer, strCountyName)
```

Once the feature is returned from the Geodatabase subsystem, the functional unit can call the Geometry subsystem to get the extent of the feature. That extent is passed to the ArcMap subsystem to change the map's extent.

Further Reading

You can find more information about application development, working with COM, and using UML in the following publications.

Eriksson, Hans-Erik and Magnus Penker, *UML Toolkit*. New York: John Wiley, 1998.

Foxall, James D., *Practical Standards for Microsoft Visual Basic*. Microsoft Corporation, 2000.

Platt, David S., *The Essence of COM*. Englewood Cliffs, NJ: Prentice-Hall PTR, 2000.

Rumbaugh, James, *Object-Oriented Modeling and Design*. Englewood Cliffs, NJ: Prentice-Hall, 1991.

CHAPTER **10**

Developing ArcGIS Applications with ArcObjects

THIS CHAPTER CONSISTS LARGELY OF A TUTORIAL that gives you a taste of developing ArcGIS applications with ArcObjects. You can complete the tutorial within two hours. The tutorial assumes that you are familiar with ArcMap commands and terminology. You also need access to the following three shape files on your system. These shape files are distributed with ArcGIS on the "ESRI Data & Maps" CD-ROMs.

- The *dtl_st.shp* shape file, containing detailed state boundaries for the United States.

- The *dtl_cnty.shp* shape file, containing detailed county boundaries for the United States.

- The *tracts.shp* shape file, containing census tract boundaries for the United States.

If you do not have these shape files on your system, copy them to an accessible directory. Keep in mind that a shape file in reality is a collection of files, so copy all files that start with *dtl_st*, *dtl_cnty*, and *tracts*. The purpose of this tutorial is to make you familiar with ArcGIS applications written in VBA. Therefore, the coding is kept simple and omits error trapping and other items you would normally consider when working with any software application.

Defining the Application

You should always have a full understanding of the application before attempting to build it. Having a full understanding means knowing what the application's requirements are. A good definition for the application provides the design and development framework. The following is a definition for the tutorial application.

The tutorial application creates a thematic layout of a county based on census tracts. You select which county and census value to map. You can also create a title for the layout, and optionally add a north arrow, scale bar, and legend.

Using the definition for guidance, you should write a stage-by-stage narrative in plain English on how the application will work. This helps you organize your development activities by revealing the user interface you need to develop and the procedures you need to write. The following narrative is for the tutorial application.

- *Stage 1:* When the application is started, it should check to make sure that required shape files are available in the map document. If not, the application should load them.

- *Stage 2:* The application creates a sorted list of state names and asks the user to select one.

- *Stage 3:* The application creates a sorted list of county names based on the selected state, and asks the user to select a county.

- *Stage 4:* The user selects the classification field and picks the number of classes.

- *Stage 5:* The user types a map title and indicates if a north arrow, scale bar, or legend should be shown on the layout.

- *Stage 6:* The application zooms to the selected county and sets a layer definition so that only the census tracts in the selected county are displayed.

- *Stage 7:* The application performs the classification based on the selected tract's field and number of classes.

- *Stage 8:* The application builds the layout using the given title and selected options.

Based on the preceding stages, you can see that you need one or more forms to collect user input, and a series of procedures to carry out the tasks.

Constructing the Application

Normally, you want to start the construction by building the user interface. Having the user interface further frames the coding process. Based on the application's definition, the user needs to provide or select the following items.

- State, county, classification field, and number of classes

- Layout title and inclusion of north arrow, scale bar, and legend

Stages 2 through 5 in the preceding application definition deal with the user interface. You can use one form to accept the user's input. However, it is recommended that you collect the user input with two forms, because the input can logically be separated into map and layout items. In any application, you need to strike a balance between the number of forms and the complexity of each form.

Building the User Interface

Figures 10-1 and 10-2 show the user interface for the application. The user selects the mapping items from the form shown in figure 10-1 and clicks on the OK button to display the form shown in figure 10-2. Next, the user provides the layout options and clicks on the OK button to complete the application. Clicking on the Cancel button ends the application. Instructions on creating the user interface follow. You create a user interface in VBA by inserting a user form and adding interface controls.

1 Start ArcMap with a new map file. In the Tools menu, select the Macros/Visual Basic Editor menu item, or alternatively press Alt-

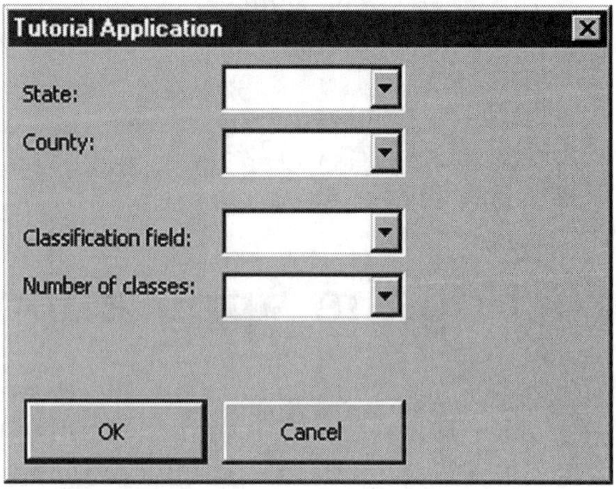

Fig. 10-1. First user form.

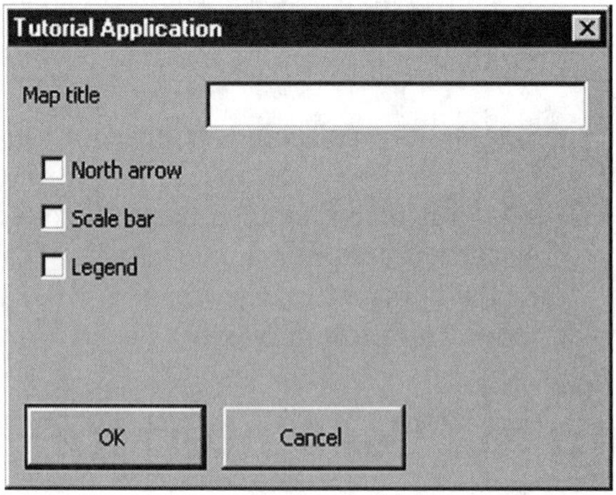

Fig. 10-2. Second user form.

F11 to open the Visual Basic Editor (VBE). Do not forget to save your map file as you complete tasks.

2   Add a new user form by selecting Insert | User Form menu item. The VBE opens a new user form and displays the toolbox, shown in figure 10-3. If you do not see the toolbox, select View | Toolbox menu to display it. The VBE identifies this new form as *UserForm1*. Change this to a more meaningful name, such as *frmClassify*, as shown in step 3.

3   Change the form name to *frmClassify* by editing the name property. If the Properties window is not displayed, select View | Properties

Fig. 10-3. Toolbox.

Window to display it. Alternatively, you can press F4. Figure 10-4 shows the Properties window for the *frmClassify* user form. Change the name by typing the new name to the right of the name property, as shown in figure 10-4.

4 Change the caption property of the user form to *Tutorial Application*.

5 Add a label to the user from by first selecting the Label control on the toolbox and then clicking on the user form. If you are not sure which icon on the toolbox represents a label, place your mouse pointer on each icon to see the tool's name. Figure 10-5 shows how

Fig. 10-4. Properties window.

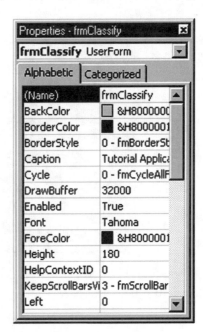

the user form appears after adding a label. While the label is selected, change its name property to *lblState* and its caption to *State:*. Then position the label as shown in figure 10-5.

Fig. 10-5. Adding a label to the user form.

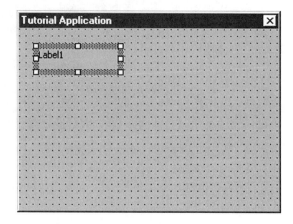

6   Add a combo box control to the user form by first selecting the Combo Box control on the toolbox and then clicking on the user form. Figure 10-6 shows how the user form appears after adding a combo box. Change the name property of the combo box to *cboState*. Change the style property to *2-fmStyleDropDownList*. Position the combo box as shown in figure 10-6.

Fig. 10-6. Adding a combo box to the user form.

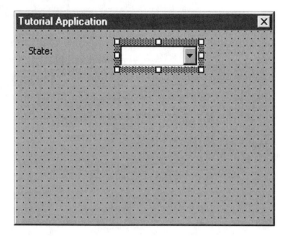

7   Add the following controls to the *frmClassify* form by repeating steps 5 and 6.

   •   Label Controls:

Name: *lblCounty*

Caption: *County:*

Name: *lblField*

Caption: *Classification field:*

Name: *lblClassCount*

Caption: *Number of classes:*

- Combo Box Controls:

Name: *cboCounty*

Style: *2-fmStyleDropDownList*

Name: *cboField*

Style: *2-fmStyleDropDownList*

Name: *cboClassCount*

Style: *2-fmStyleDropDownList*

8 Add the OK and Cancel buttons. Click on the Command button on the toolbox and click on the user form for each button. Name the buttons *cmdOK* and *cmdCancel*, and set their caption properties to OK and Cancel, respectively. Set the default property of the OK button to True, so that pressing the Enter key while the application is running will be the same as clicking on the OK button. Set the cancel property of the Cancel button to True, so that pressing the Escape key while the application is running will be the same as clicking on the Cancel button. Arrange the buttons as shown in figure 10-1.

9 Create a second user form, based on the following properties, to match that shown in figure 10-2.

- User Form:

Name: *frmLayout*

Caption: *Tutorial Application*

- Label Control:

Name: *lblTitle*

Caption: *Map control*

- Text Box Control:

  Name: *txtTitle*
- Check Box Controls:

  Name: *chkNorthArrow*

  Caption: *North arrow*

  Name: *chkScale*

  Caption: *Scale bar*

  Name: *chkLegend*

  Caption: *Legend*
- Command Buttons:

  Name: *cmdOK*

  Caption: *OK*

  Name: *cmdCancel*

  Caption: *Cancel*

The name property is very important because your application accesses user controls by their names. Therefore, make sure you have set all name properties correctly. You have now completed creating the user interface. Save your work. Next, you need to determine the required procedures and add the code to complete the application.

Building the Procedures

Procedures create the application logic and process. You need to determine the required procedures based on the application's definition. For example, stage 1 of the application's definition suggests that the application should load the required shape files if they do not exist in the map document. Therefore, one procedure could check for the existence and loading of the shape files. This tutorial divides the application logic into the following procedures.

- Procedure *Tutorial*: This is the starting point for the application. It checks for the existence of data, loads shape files if needed, and displays the user interface.

- Procedure *GetLayer*: The *Tutorial* procedure calls this routine to access a map layer.

- Procedure *AddShapeFile*: The *Tutorial* procedure calls this routine to load a shape file as a map layer.

- Procedure *PopulateStateCombo*: This procedure creates a sorted list of states, and populates the state combo box control as described in stage 2 of the application's definition.

- Procedure *cboState_Change*: This procedure is executed each time the user selects a different state. It populates the county combo box for the selected state per stage 3 of the application's definition.

- Procedure *PopulateClassificationCombo*: As its name suggests, this procedure populates the classification field combo box with a list of fields that can be used to create a thematic map.

- Procedure *PopulateClassCountCombo*: In stage 4 of the application's definition, the user can select the classification count and number of classes. This procedure populates the combo box for the number of classes.

- Procedure *ZoomToCounty*: This procedure changes the map extent to the extent of the selected county, as required by stage 6 of the application's definition. It also sets a layer definition so that only the census tracts of the selected county are displayed.

- Procedure *ClassifyMap*: Tracts are classified based on the user input in this procedure, as described in stage 7 of the application's definition.

- Procedure *BuildLayout*: In the final step of the application, this procedure builds the map layout based on user input.

Procedures can be created in several places in the VBA environment. In the following, you will place the procedures that start the application in a module, and the rest in their respective forms. As stated at the beginning of the chapter, to keep the coding simple, considerations such as modularization and code reuse have been omitted. One tactic is to place very little code in the forms and try to take advantage of class modules. However, addressing such concepts is beyond the scope of this book. See the "Further Reading" section at the end of Chapter 9 for a list of books on these topics.

The procedures are available for download through the publisher's web site. See the Introduction to this book for instructions on how to download the procedures. You will note in the following that the procedures are broken down into logical steps. Each step performs a task.

1 Create a new module by selecting Insert | Module. Display the property window for this module by pressing F4. Change the module name to *Main*.

2 Add the following code to the module *Main*.

CODE VBA10-1

```
Option Explicit
Global g_pStateLayer As IFeatureLayer
Global g_pCountyLayer As IFeatureLayer
Global g_pTractLayer As IFeatureLayer
Public Sub Tutorial()
  ' This procedure is called when user clicks on the
  ' customized button
  '
  ' (1) Edit the following constants to match your
  '     environment
  Const c_strDataPath = "E:\arcgis\arcdata\cd3\usa\"
  Const c_strStateFileName = "dtl_st.shp"
  Const c_strStateLayerName = "State"
  Const c_strCountyFileName = "dtl_cnty.shp"
  Const c_strCountyLayerName = "County"
  Const c_strTractFileName = "tracts.shp"
  Const c_strTractLayerName = "Census Tract"
  '
  ' (2) Load the shape files if necessary
  Dim pLayer As IFeatureLayer
  ' State shape file
  Set pLayer = GetLayer(c_strStateLayerName)
  If pLayer Is Nothing Then
    Set pLayer = AddShapeFile(c_strDataPath, _
    c_strStateFileName, c_strStateLayerName)
    If pLayer Is Nothing Then
      MsgBox "Unable to locate " & c_strDataPath & _
      c_strStateFileName & " shape file."
      Exit Sub
    End If
  End If
```

```
      End If
      Set g_pStateLayer = pLayer
      ' County shape file
      Set pLayer = GetLayer(c_strCountyLayerName)
      If pLayer Is Nothing Then
        Set pLayer = AddShapeFile(c_strDataPath, _
        c_strCountyFileName, c_strCountyLayerName)
        If pLayer Is Nothing Then
          MsgBox "Unable to locate " & c_strDataPath & _
          c_strCountyFileName & " shape file."
          Exit Sub
        End If
        ' Make county invisible
        pLayer.Visible = False
      End If
      Set g_pCountyLayer = pLayer
      ' Census tract shape file
      Set pLayer = GetLayer(c_strTractLayerName)
      If pLayer Is Nothing Then
        Set pLayer = AddShapeFile(c_strDataPath, _
        c_strTractFileName, c_strTractLayerName)
        If pLayer Is Nothing Then
          MsgBox "Unable to locate " & c_strDataPath & _
          c_strTractFileName & " shape file."
          Exit Sub
        End If
        ' Make tract invisible
        pLayer.Visible = False
      End If
      Set g_pTractLayer = pLayer
      '
      ' (3) Display the user interface form and populate
      '    its combo boxes
      frmClassify.PopulateClassCountCombo
      frmClassify.PopulateClassificationCombo
      frmClassify.PopulateStateCombo
      frmClassify.Show
    End Sub
    Private Function GetLayer(strLayerName As String) As _
    IFeatureLayer
      ' This function accepts a layer name and returns
```

```
' the layer if available, otherwise returns "Nothing".
'
' (1) Access the document's map
Dim pMxDoc As IMxDocument
Dim pMap As IMap
Set pMxDoc = Application.Document
Set pMap = pMxDoc.FocusMap
'
' (2) Search through layers for the given layer name
Dim lngIndex As Long
Set GetLayer = Nothing
For lngIndex = 0 To pMap.LayerCount - 1
  If pMap.Layer(lngIndex).Name = strLayerName Then
    Set GetLayer = pMap.Layer(lngIndex)
    Exit For
  End If
Next lngIndex
End Function
Private Function AddShapeFile(strPath As String, _
strFile As String, strName As String) As IFeatureLayer
  ' This function adds the specified shapefile and
  ' returns the layer. It returns "Nothing" if not
  ' successful.
  '
  ' (1) Make sure the shape file exists
  If Len(Dir(strPath & strFile)) = 0 Then
    ' File does not exist
    Set AddShapeFile = Nothing
    Exit Function
  End If
  '
  ' (2) Create a workspace to represent the datasource
  Dim pWorkspaceFactory As IWorkspaceFactory
  Dim pFeatureWorkspace As IFeatureWorkspace
  Set pWorkspaceFactory = New ShapefileWorkspaceFactory
  Set pFeatureWorkspace = _
  pWorkspaceFactory.OpenFromFile(strPath, 0)
  '
  ' (3) Access the shape file through a feature layer
  Dim pClass As IFeatureClass
```

```
    Dim pFeatureLayer As IFeatureLayer
    Set pClass = pFeatureWorkspace.OpenFeatureClass(strFile)
    Set pFeatureLayer = New FeatureLayer
    Set pFeatureLayer.FeatureClass = pClass
    pFeatureLayer.Name = strName
    '
    ' (4) Add layer to the map
    Dim pMxDoc As IMxDocument
    Dim pMap As IMap
    Set pMxDoc = Application.Document
    Set pMap = pMxDoc.FocusMap
    pMap.AddLayer pFeatureLayer
    Set AddShapeFile = pFeatureLayer
End Function
```

3 Change the value of *c_strDataPath*, as it appears in paragraph 1 of the *Tutorial* procedure, to your directory where you have saved the shape files.

4 Open the user form *frmClassify* and display its code editor by selecting View | Code. Alternatively, you can press F7. Then add the following procedures.

CODE
VBA10-2

```
Option Explicit
Public Sub PopulateStateCombo()
    ' This procedure uses the State layer to populate
    ' cboState with state names. The State layer is available
    ' through a global variable that was set in module Main.
    '
    ' (1) Access the State layer
    Dim pLayer As IFeatureLayer
    Dim pFeatureClass As IFeatureClass
    Set pLayer = Main.g_pStateLayer
    Set pFeatureClass = pLayer.FeatureClass
    '
    ' (2) Sort State layer's table
    Dim pTableSort As ITableSort
    Set pTableSort = New TableSort
    pTableSort.Fields = "STATE_NAME"
    pTableSort.Ascending("STATE_NAME") = True
    Set pTableSort.Table = pFeatureClass
```

```
  pTableSort.Sort Nothing
  '
  ' (3) Populate the State's combo box
  Dim pCursor As ICursor
  Dim pRow As IRow
  Dim lngFieldIndex As Long
  Set pCursor = pTableSort.Rows
  lngFieldIndex = pFeatureClass.FindField("STATE_NAME")
  Set pRow = pCursor.NextRow
  Do While Not pRow Is Nothing
    cboState.AddItem pRow.Value(lngFieldIndex)
    Set pRow = pCursor.NextRow
  Loop
  cboState.ListIndex = 0
End Sub
Public Sub PopulateClassificationCombo()
  ' The Classification field combo always has a list
  ' of numeric fields in the census tract layer. This
  ' procedure populates it.
  '
  ' (1) Access the Tract layer
  Dim pLayer As IFeatureLayer
  Dim pFeatureClass As IFeatureClass
  Set pLayer = Main.g_pTractLayer
  Set pFeatureClass = pLayer.FeatureClass
  '
  ' (2) Populate combo box with numeric field names
  Dim pFields As IFields
  Dim pField As IField
  Dim lngIndex As Long
  Set pFields = pFeatureClass.Fields
  For lngIndex = 0 To pFields.FieldCount - 1
    Set pField = pFields.Field(lngIndex)
    If pField.Type = esriFieldTypeDouble Or _
    pField.Type = esriFieldTypeInteger Or _
    pField.Type = esriFieldTypeSingle Then
      cboField.AddItem pField.AliasName
    End If
  Next lngIndex
  cboField.ListIndex = 0
```

```
End Sub
Public Sub PopulateClassCountCombo()
  ' The application provides from 2 to 9 classes.
  ' This procedure populates the class count choices
  ' in the class count combo box.
  Dim lngIndex As Long
  For lngIndex = 2 To 9
    cboClassCount.AddItem lngIndex
  Next lngIndex
  cboClassCount.ListIndex = 0
End Sub
Private Sub cboState_Change()
  ' This procedure is executed each time a different
  ' state is selected. It populates the county combo
  ' box based on the selected state.
  '
  ' (1) Access the County layer
  Dim pLayer As IFeatureLayer
  Dim pFeatureClass As IFeatureClass
  Set pLayer = Main.g_pCountyLayer
  Set pFeatureClass = pLayer.FeatureClass
  '
  ' (2) Query and sort County layer's table
  Dim pTableSort As ITableSort
  Dim pQueryFilter As IQueryFilter
  Set pQueryFilter = New QueryFilter
  'pQueryFilter.SubFields = "STATE_NAME,NAME"
  pQueryFilter.WhereClause = "STATE_NAME = '" & _
  cboState.Text & "'"
  Set pTableSort = New TableSort
  pTableSort.Fields = "STATE_NAME,NAME"
  pTableSort.Ascending("STATE_NAME") = True
  pTableSort.Ascending("NAME") = True
  Set pTableSort.QueryFilter = pQueryFilter
  Set pTableSort.Table = pFeatureClass
  pTableSort.Sort Nothing
  '
  ' (3) Populate the County's combo box
  Dim pCursor As ICursor
  Dim pRow As IRow
```

```
Dim lngFieldIndex As Long
' Clear any existing county names
cboCounty.Clear
Set pCursor = pTableSort.Rows
lngFieldIndex = pFeatureClass.FindField("NAME")
Set pRow = pCursor.NextRow
Do While Not pRow Is Nothing
  cboCounty.AddItem pRow.Value(lngFieldIndex)
  Set pRow = pCursor.NextRow
Loop
cboCounty.ListIndex = 0
End Sub
Private Sub cmdCancel_Click()
  End
End Sub
Private Sub cmdOK_Click()
  frmClassify.Hide
  frmLayout.Show
End Sub
```

5 Open the user form *frmLayout* and display its code editor by select-
ing View | Code or by pressing F7. Then add the following proce-
dures.

CODE
VBA10-3

```
Option Explicit
Private Sub ZoomToCounty()
  ' This procedure sets the definition for the Tracts layer
  ' so that only tracts of the selected county is displayed.
  ' It also zooms to the selected county.
  '
  ' (1) Collect the required information.
  Dim strStateName As String
  Dim strCountyName As String
  Dim pTractLayer As IFeatureLayer
  Dim pCountyLayer As IFeatureLayer
  Dim pStateLayer As IFeatureLayer
  strStateName = frmClassify.cboState
  strCountyName = frmClassify.cboCounty
  Set pTractLayer = Main.g_pTractLayer
  Set pCountyLayer = Main.g_pCountyLayer
  Set pStateLayer = Main.g_pStateLayer
```

```
'
' (2) Get the FIPS values for the state and county
'   and zoom to the county.
Dim pQueryFilter As IQueryFilter
Dim pFeatureCursor As IFeatureCursor
Dim pFields As IFields
Dim pField As IField
Dim lngIndex As Long
Dim pFeature As IFeature
Dim strStateFIPS As String
Dim strCountyFIPS As String
Dim pMxDocument As IMxDocument
Dim pMap As IMap
Dim pActiveView As IActiveView
Dim pCountyEnvelope As IEnvelope
' State FIPS
Set pQueryFilter = New QueryFilter
pQueryFilter.WhereClause = "STATE_NAME = '" & _
strStateName & "'"
Set pFeatureCursor = pStateLayer.Search(pQueryFilter, False)
lngIndex = pFeatureCursor.FindField("STATE_FIPS")
Set pFeature = pFeatureCursor.NextFeature
If pFeature Is Nothing Then
  MsgBox "Unable to find State FIPS", _
  vbCritical + vbOKOnly
  Exit Sub
End If
strStateFIPS = pFeature.Value(lngIndex)
' County FIPS
Set pQueryFilter = New QueryFilter
pQueryFilter.WhereClause = "STATE_FIPS = '" & strStateFIPS & _
"' and NAME = '" & strCountyName & "'"
Set pFeatureCursor = pCountyLayer.Search(pQueryFilter, False)
lngIndex = pFeatureCursor.FindField("CNTY_FIPS")
Set pFeature = pFeatureCursor.NextFeature
If pFeature Is Nothing Then
  MsgBox "Unable to find County FIPS", _
  vbCritical + vbOKOnly
  Exit Sub
End If
```

```
      strCountyFIPS = pFeature.Value(lngIndex)
      ' Zoom to the county
      Set pCountyEnvelope = pFeature.Extent
      Set pMxDocument = Application.Document
      Set pMap = pMxDocument.FocusMap
      Set pActiveView = pMap
      pActiveView.Extent = pCountyEnvelope.Envelope
      '
      ' (3) Set the definition for the Tract layer to
      '    only display tracts of the selected county.
      '    Also refresh the displayed view.
      Dim pFeatureLayerDefinition As IFeatureLayerDefinition
      Set pFeatureLayerDefinition = pTractLayer
      pFeatureLayerDefinition.DefinitionExpression = _
      "STCOFIPS = '" & _
      strStateFIPS & strCountyFIPS & "'"
      pTractLayer.Visible = True
      pCountyLayer.Visible = False
      pStateLayer.Visible = False
      pActiveView.Refresh
End Sub
Private Sub ClassifyMap()
   ' This procedure classifies the tract layer
   ' based on the selections on frmClassify.
   '
   ' (1) Collect the required information
   Dim strClassFieldName As String
   Dim lngClassCount As Long
   Dim pLayer As IFeatureLayer
   Dim pFeatureClass As IFeatureClass
   strClassFieldName = frmClassify.cboField
   lngClassCount = frmClassify.cboClassCount
   Set pLayer = Main.g_pTractLayer
   Set pFeatureClass = pLayer.FeatureClass
   '
   ' (2) Use a histogram to calculate class breaks
   Dim pTable As ITable
   Dim pTableHistogram As ITableHistogram
   Dim pHistogram As IHistogram
```

```
Dim vntDataValues As Variant
Dim vntDataFrequencies As Variant
Set pTable = pLayer
Set pTableHistogram = New TableHistogram
Set pTableHistogram.Table = pTable
pTableHistogram.Field = strClassFieldName
Set pHistogram = pTableHistogram
pHistogram.GetHistogram vntDataValues, vntDataFrequencies
'
' (3) Set up a classification
Dim pClassify As IClassify
Dim dblBreakValues() As Double
Set pClassify = New Quantile
pClassify.SetHistogramData vntDataValues, vntDataFrequencies
pClassify.Classify lngClassCount
dblBreakValues = pClassify.ClassBreaks
'
' (4) Build a color ramp for renderer
Dim pColorRamp As IColorRamp
Dim blnOK As Boolean
Set pColorRamp = New RandomColorRamp
pColorRamp.Size = lngClassCount
pColorRamp.CreateRamp blnOK
'
' (4) Create a ClassBreaks type renderer
Dim pRenderer As IClassBreaksRenderer
Dim lngIndex As Long
Dim pColor As IColor
Dim pEnumColors As IEnumColors
Dim pFillSymbol As IFillSymbol
Set pRenderer = New ClassBreaksRenderer
pRenderer.Field = strClassFieldName
pRenderer.BreakCount = lngClassCount
pRenderer.MinimumBreak = dblBreakValues(0)
For lngIndex = 0 To pRenderer.BreakCount - 1
  Set pColor = pColorRamp.Color(lngIndex)
  Set pFillSymbol = New SimpleFillSymbol
  pFillSymbol.Color = pColor
  pRenderer.Symbol(lngIndex) = pFillSymbol
  pRenderer.Break(lngIndex) = dblBreakValues(lngIndex + 1)
```

```
      If lngIndex = 0 Then
        pRenderer.Label(lngIndex) = "0 - " & _
        dblBreakValues(lngIndex + 1)
      Else
        pRenderer.Label(lngIndex) = dblBreakValues(lngIndex) _
        & " - " & dblBreakValues(lngIndex + 1)
      End If
    Next lngIndex
    '
    ' (5) Update the legend and display the thematic map
    Dim pGeoFeaturelayer As IGeoFeatureLayer
    Dim pDoc As IMxDocument
    Dim pLegendInfo As ILegendInfo
    Set pLegendInfo = pRenderer
    pLegendInfo.LegendGroup(0).Heading = strClassFieldName
    Set pGeoFeaturelayer = pLayer
    Set pGeoFeaturelayer.Renderer = pRenderer
    Set pDoc = ThisDocument
    pDoc.UpdateContents
    pDoc.ActivatedView.Refresh
End Sub
Private Sub BuildLayout()
    ' This procedure creates the layout based on
    ' user's options.
    '
    ' (1) Set the page size
    Dim pMxDocument As IMxDocument
    Dim pPageLayout As IPageLayout
    Dim pPage As IPage
    Dim pActiveView As IActiveView
    Dim dblPageWidth As Double, dblPageHeight As Double
    Set pMxDocument = Application.Document
    Set pActiveView = pMxDocument.PageLayout
    Set pPageLayout = pMxDocument.PageLayout
    Set pPage = pPageLayout.Page
    pPage.Units = esriInches
    dblPageWidth = 8.5
    dblPageHeight = 11
    pPage.PutCustomSize dblPageWidth, dblPageHeight
    '
```

```
' (2) Configure location of the map frame
Dim pMap As IMap
Dim pGraphicsContainer As IGraphicsContainer
Dim pMapFrame As IMapFrame
Dim pElement As IElement
Dim pEnvelope As IEnvelope
Set pMap = pMxDocument.FocusMap
Set pGraphicsContainer = pPageLayout
Set pMapFrame = pGraphicsContainer.FindFrame(pMap)
Set pElement = pMapFrame
Set pEnvelope = New Envelope
pEnvelope.XMin = 0.5
pEnvelope.XMax = 8
pEnvelope.YMin = 4
pEnvelope.YMax = 10.5
pElement.Geometry = pEnvelope
'
' (3) Add text and other layout elements.
Dim pTextElement As ITextElement
Dim pTextSymbol As ITextSymbol
Dim pFont As IFontDisp
Dim pID As UID
Dim pMapSurround As IMapSurround
Dim pMapSurroundFrame As IMapSurroundFrame
' Text
If frmLayout.txtTitle > "" Then
  Set pTextElement = New TextElement
  pTextElement.Text = frmLayout.txtTitle
  Set pTextSymbol = pTextElement.Symbol
  Set pFont = pTextSymbol.Font
  pFont.Size = 24
  pTextSymbol.Font = pFont
  pTextElement.Symbol = pTextSymbol
  Set pElement = pTextElement
  Set pEnvelope = New Envelope
  pEnvelope.XMin = 0.5
  pEnvelope.XMax = 8
  pEnvelope.YMin = 3
  pEnvelope.YMax = 3.75
  pElement.Geometry = pEnvelope
```

```
      pGraphicsContainer.AddElement pTextElement, 0
End If
' North Arrow
If frmLayout.chkNorthArrow Then
  Set pEnvelope = New Envelope
  pEnvelope.XMin = 5
  pEnvelope.XMax = 8
  pEnvelope.YMin = 2
  pEnvelope.YMax = 2.75
  Set pID = New UID
  pID.Value = "esriCore.MarkerNorthArrow"
  Set pMapSurroundFrame = _
  pMapFrame.CreateSurroundFrame(pID, Nothing)
  pMapSurroundFrame.MapSurround.Name = "North Arrow"
  Set pElement = pMapSurroundFrame
  pElement.Geometry = pEnvelope
  pElement.Activate pActiveView.ScreenDisplay
  pGraphicsContainer.AddElement pElement, 0
End If
' Scale
If frmLayout.chkScale Then
  Set pEnvelope = New Envelope
  pEnvelope.XMin = 5
  pEnvelope.XMax = 8
  pEnvelope.YMin = 1
  pEnvelope.YMax = 1.75
  Set pID = New UID
  pID.Value = "esriCore.ScaleLine"
  Set pMapSurroundFrame = _
  pMapFrame.CreateSurroundFrame(pID, Nothing)
  pMapSurroundFrame.MapSurround.Name = "Scale Line"
  Set pElement = pMapSurroundFrame
  pElement.Geometry = pEnvelope
  pElement.Activate pActiveView.ScreenDisplay
  pGraphicsContainer.AddElement pElement, 0
End If
' Legend
If frmLayout.chkLegend Then
  Set pEnvelope = New Envelope
  pEnvelope.XMin = 0.5
```

```
      pEnvelope.XMax = 3.5
      pEnvelope.YMin = 1
      pEnvelope.YMax = 2.75
      Set pID = New UID
      pID.Value = "esriCore.Legend"
      Set pMapSurroundFrame = _
      pMapFrame.CreateSurroundFrame(pID, Nothing)
      pMapSurroundFrame.MapSurround.Name = "Legend"
      Set pElement = pMapSurroundFrame
      pElement.Geometry = pEnvelope
      pElement.Activate pActiveView.ScreenDisplay
      pGraphicsContainer.AddElement pElement, 0
   End If
   '
   ' (4) Unselect the layout elements and refresh the map
   Dim pGraphicsContainerSelect As IGraphicsContainerSelect
   Set pGraphicsContainerSelect = pPageLayout
   pGraphicsContainerSelect.UnselectAllElements
   pActiveView.Refresh
End Sub
Private Sub cmdCancel_Click()
   End
End Sub
Private Sub cmdOK_Click()
   ' This procedure calls other routines to create
   ' the layout.
   '
   ' (1) Change the map extent
   ZoomToCounty
   '
   ' (2) Create the thematic map
   ClassifyMap
   '
   ' (3)
   BuildLayout
   End Sub
```

6 Save your work.

You have added two other procedures, named *cmdCancel_Click* and *cmdOK_Click*, to the forms. These procedures execute when the user clicks on the Cancel or the OK button. You have now finished constructing the tutorial application. To run this application from ArcMap, you need to create a customized button, as explained in the following section.

Running the Application

You can run the application from the VBE by opening the Main module and pressing F5, or by selecting Run | Run Sub. This method of running the application is fine while developing and testing it. It is more appropriate, however, to create a customized button, as explained in the following, and run the application when the button is clicked.

1   Make the ArcMap window active. Select Tools | Customize to open the Customize dialog box, shown in figure 10-7. Click on the Toolbars tab.

Fig. 10-7. Customize dialog box and Toolbars tab.

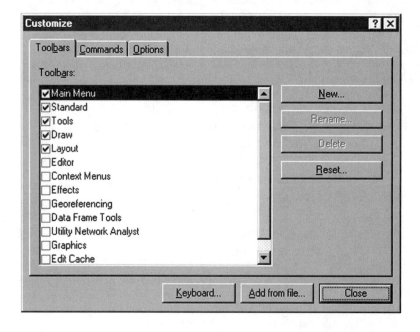

2   In the Toolbars tab, click on the New button to display the New Toolbar dialog box. Click on OK to save the new toolbar in the current document. Make sure the new toolbar is visible. If it is not, mark the corresponding check box under the Toolbars tab.

3   Click on the Commands tab of the Customize dialog box, and select the Macros category. Select *Project.Main.Tutorial* from the commands list, shown in figure 10-8.

Fig. 10-8. Customize dialog box and Commands tab.

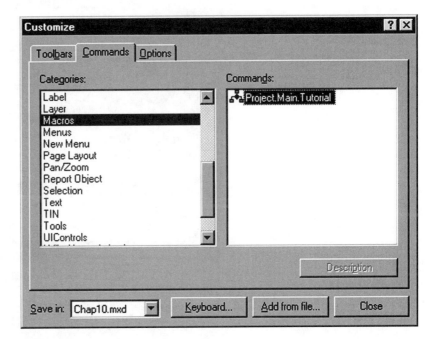

4   Drag the *Project.Main.Tutorial* command from the Customize dialog box to the new toolbar you just created. The command appears with a default icon on your toolbar.

5   Close the Customize dialog box by clicking on the Close button. Save your work.

You can now run the application by clicking on the button in your new toolbar. Figures 10-9 through 10-11 show how the application is executed, and its result.

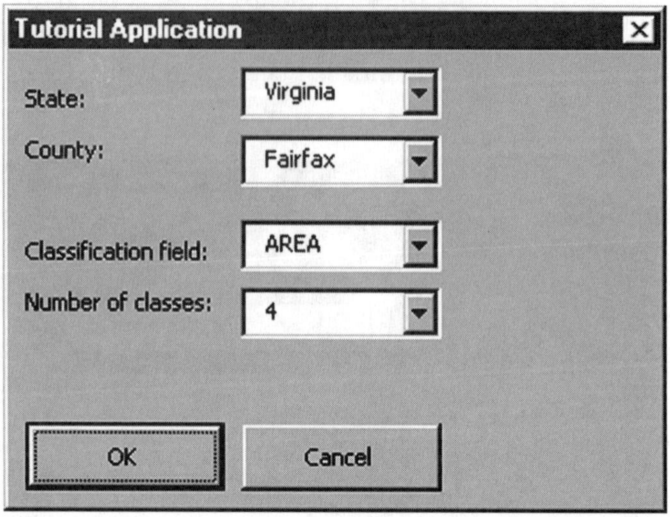

Fig. 10-9. Application's classify form.

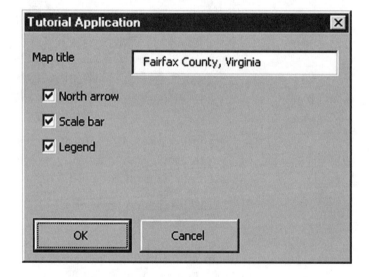

Fig. 10-10. Application's layout form.

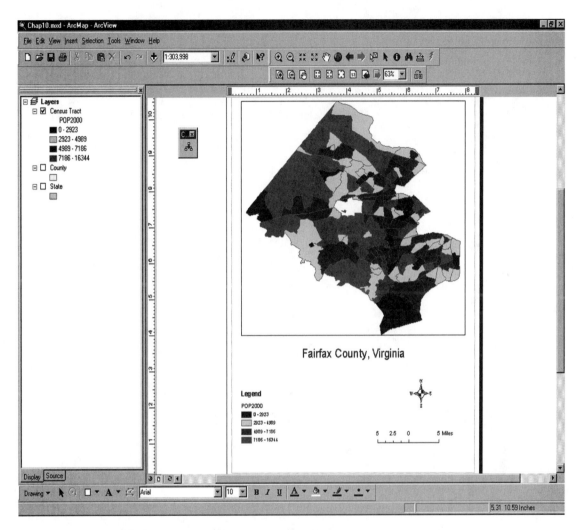

Fig. 10-11. Application's result.

Expanding the Application

If you run the tutorial application more than once, you notice that the layout items such as title, scale bar, and so on multiply because the old ones are not removed when the new ones are added. As an exercise, try adding the required code to remove the old layout items before adding new ones.

11
Customizing the ArcMap User Interface

YOUR APPLICATION IS ACCESSED AND EXECUTED through its user interface. The user interface elements consist of basic controls (such as menu items, toolbars, and dialog boxes) and more advanced controls, such as user forms. You can customize the ArcMap user interface to create menus or tool buttons for your application. You can also work with the ArcMap status bar, change the cursor's shape, and interact with your application's user interface through dialog boxes.

This chapter shows you how to customize ArcMap menus and toolbars interactively or through your application. It also shows you how to manipulate the status bar, change the mouse cursor's shape, and work with dialog boxes inside your application.

What Can Be Customized

There are many ways to alter the look and behavior of the ArcMap user interface controls. Because ArcMap is based on Microsoft's Component Object Model (COM) technology, you customize by identifying and working with the interface of the objects that form the user interface.

Figure 11-1 shows the Main menu and the Tools and Standard toolbars. The Main menu and Standard toolbar appear in ArcMap by default. A toolbar can display its menu items as they appear under the Main menu, or it can combine menus, buttons, tools, combo boxes, and edit boxes.

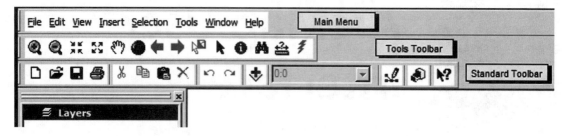

Fig. 11-1. Menu and toolbars.

Figure 11-2 shows the status bar. ArcMap uses the status bar to provide information such as the selected command. Your application can also use the status bar to display text messages or show a progress bar.

 TIP: *A progress bar can reduce user frustration with processes that take a long time.*

*Fig. 11-2.
Status bar.*

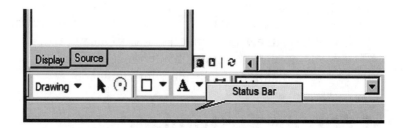

In an application, the user interface is not limited to controls such as menus and tools. Often your VBA macro needs to interact with its user to get required information. You can create your own user forms in the VBA environment to satisfy this need. You can also use the various dialog boxes provided by ArcObjects. If appropriate, using these dialog boxes is more efficient than developing your own user forms.

Interactive Customization

Toolbars in ArcMap hold the user interface controls that are also known as controls. Toolbars can have the following five types of commands.

- You can arrange commands into menu controls. Menu controls organize related commands in a list. ArcMap also offers pop-up menus that appear by pressing the right-hand mouse button. The pop-up menus are also known as context menus.

- Buttons and menu items are individual controls that appear on a toolbar or inside a menu list. You can perform a task or run a script by clicking on them.

- Tool commands perform a task or run a script when you interact with the map display. You first select the tool and then apply it to the map display.

- You can select an option from a combo box control. You can also type in the option's value if applicable.

- The users of your application can enter a value in the text box controls, also referred to as edit boxes.

You can customize the existing toolbars and controls, or you can create your own toolbars and controls. This customization is done through the Customize dialog box, shown in figure 11-3. To open the Customize dialog box, select Tools | Customize.

 TIP: *You can open the Customize dialog by double clicking on an empty area of a toolbar, or by pressing the right-hand mouse button on a toolbar and selecting Context | Customize.*

At the bottom of the Customize dialog you see three buttons: Keyboard, *Add from file*, and Close. The Close button dismisses the Customize dialog box. The *Add from file* button facilitates adding a custom command built outside VBA (such as a Visual Basic or C + + DLL). Non-VBA custom commands are outside the scope of this book.

Fig. 11-3. Customize dialog box showing Toolbars tab.

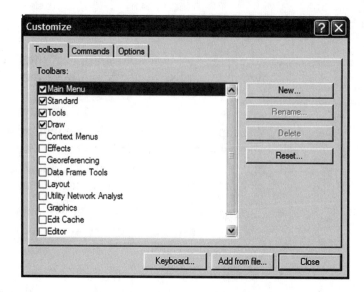

The Keyboard button customizes the keyboard by managing the commands' shortcuts. Selecting this button displays the Customize Keyboard dialog box, shown in figure 11-4.

Fig. 11-4. Customize Keyboard dialog box.

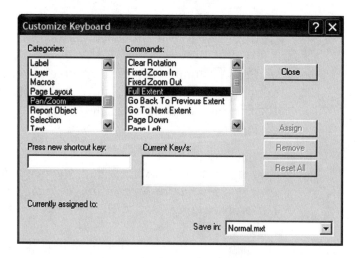

The following steps show you how to assign Ctrl + Alt + F to the Full Extent command.

1   Open the Customize Keyboard dialog by clicking on the Keyboard button in the Customize dialog box.

2 Select Pan | Zoom from the Categories list, and then select Full Extent from the Commands list.

3 Place the cursor in the text box labeled *Press new shortcut key*, and then press Ctrl + Alt + F. This action adds the *Ctrl + Alt + F* text to the *Press new shortcut key* text box.

4 Pay attention to the value of the *Currently assigned to* text. It indicates whether or not the shortcut is already used by another command. Assigning a shortcut that is already used overrides the original assignment.

5 Click on the Assign button to place the *Ctrl + Alt + F* text in the Current Keys list. You can have more than one shortcut assigned to a command.

6 Click on the Close button in the Customize Keyboard dialog and then close the Customize dialog box.

7 Test your new shortcut by zooming in to your map and then pressing the Ctrl + Alt + F keys to return to full extent.

 TIP: *Saving the shortcut or any customization in the Normal template, as shown in figure 11-4, automatically loads the customization each time you start ArcMap. See the "Saving Your Customizations" section in this chapter.*

The Customize dialog box, shown in figure 11-3, has three tabs: Toolbars, Commands, and Options. Each tab is explained in the following sections.

Toolbars Tab

Select the Toolbars tab to see a list of available toolbars, shown in figure 11-3. A check mark next to a toolbar indicates its visibility. A toolbar is displayed when there is a check mark, and it is hidden when there is no check mark.

Using the buttons on this tab, you can create new toolbars, rename or remove toolbars, and reset toolbars to their default settings. You cannot rename or delete the built-in toolbars.

Commands Tab

The Commands tab assists you in adding, removing, and rearranging the commands on any toolbar. The following steps take you through the process of creating a new toolbar with selection-related commands.

1   Open the Customize dialog box as described previously.

2   In the Toolbars tab, click on the New button to display the New Toolbar dialog. In this dialog, type *Selection* in the Toolbar Name field and then click on the OK button. This action creates a new, empty toolbar.

3   Click on the Commands tab in the Customize dialog, shown in figure 11-5.

Fig. 11-5. Customize dialog box showing the Commands tab.

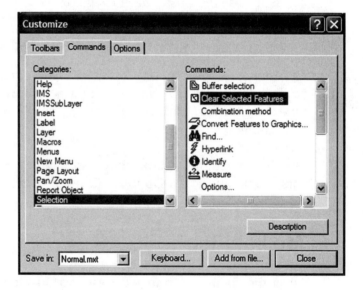

4   Select the Selection option from the Categories list to display selection-related commands.

5   Drag the Clear Selected Feature command from the Commands list to the new, empty toolbar you created in step 2. This action adds the command to the toolbar. Add the Select All, Select Features, Switch Selection, and Zoom To Selected Features commands by dragging them from the Commands list onto the new toolbar. Figure 11-6 shows the new toolbar.

6 Close the Customize dialog.

Fig. 11-6. New toolbar.

You can rearrange or remove tools from a toolbar while the Customize dialog is open. To delete a tool, drag and release it outside the ArcMap window.

Options Tab

The Options tab of the Customize dialog helps you set the following options.

- Increase the size of icons. This has no effect on the menus.

- Turn on or off the display of tool tips.

- Set the default location for saving your customization.

The preceding options are presented in the form of check boxes on the Options tab, as shown in figure 11-7. The Options tab contains the following three pushbuttons.

- *Lock Customization:* Restricts access to the Customize dialog by requiring a password.

- *Change VBA Security:* Helps you set a protection level against computer viruses that can be distributed through a VBA macro.

- *Selecting Upgrade ArcID Module:* Updates a file that allows you refer to added COM objects by name. This option is beyond the scope of this book.

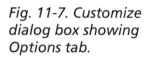

Fig. 11-7. Customize dialog box showing Options tab.

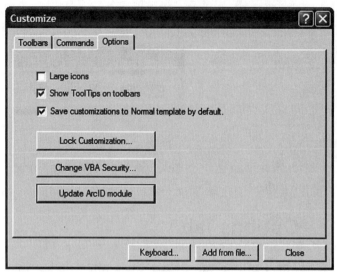

Saving Your Customization

You can save your customization of the interface if you made the changes interactively. Customizations done through a VBA macro are temporary for the current ArcMap session and cannot be saved. You can save the interactive changes to the user interface in three different manners: save to the current map document, create a template, or save to the Normal template. When you save to the current map, your customization appears only when that specific map document is open.

You can save your customization to a template, and when you create a new map document you can base it on that template. In this manner a new map document can have your customized user interface. By default, new map documents are based on the Normal template. Saving your customization in the Normal template ensures that all new map documents have your customized user interface.

Before making any changes to the user interface, decide where to save the customization. You do this via the Commands tab of the Customize dialog, shown in figure 11-5. The Commands tab contains a *Save in* combo box that sets how your customization is saved. By default, the value of the *Save in* combo box is set to the Normal template. If you wish to save your customization in the current map document, select the map document from the combo box before making your changes.

The following steps take you through the process of saving your customization in a template.

1 Start a new map document.

2 Save your new map document as an ArcMap template.

3 Set the *Save in* combo box of the Commands tab on the Customize dialog to the current map document.

4 Make your changes to the user interface.

5 Save your template again.

ArcMap automatically creates the Normal template and places it in your *Profiles* directory, which on the Windows operating system is in one of the following folders.

- On Windows NT: *C:\WINNT\Profiles\< your username > \Application Data\ESRI\ArcMap\Templates*

- On Windows 2000: *C:\Documents and Settings\< your username > \Application Data\ESRI\ArcMap\Templates*

 TIP: *You can remove all customizations to the Normal template by deleting this template from your* Profiles *directory. The next time you start ArcMap, ArcGIS will create a new Normal template.*

Programming the User Interface

You can modify the user interface programmatically. Keep in mind that customizations done by VBA macros are temporary and cannot be saved. Often you may combine the interactive and programmatic customization. For example, you may interactively create a new menu to start your VBA macro that further customizes the user interface. You need VBA macros if you wish to utilize user forms.

The following steps take you through the process of creating a VBA macro that displays the current date and time, as well as a custom command for running the macro. Start by opening the VBA editor (Tools | Macros | Visual Basic Editor). The shortcut key combination for the VBA editor is Alt + F11.

1 On the Project Explorer window of the VBA editor (see figure 11-8), double click on *ThisDocument* to open the Code window (see figure 11-9).

2 Add the following VBA code to the Code window.

```
Public Sub ShowTime()
  MsgBox Now()
End Sub
```

3 Save your map document.

Fig. 11-8. VBA editor showing Project Explorer.

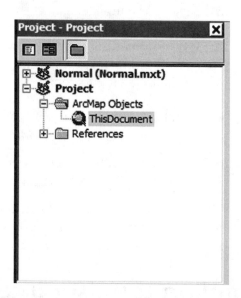

Fig. 11-9. VBA editor showing the Code window.

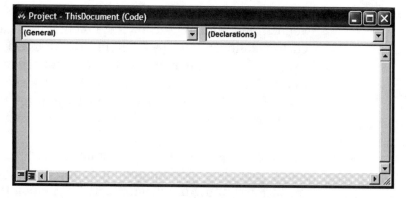

You now need to create a new toolbar and place a new command button on the toolbar. The following steps take you through this process.

1   Open the Customize dialog box as described previously.

2   On the Toolbars tab, click on the New button to display the New Toolbar dialog. In this dialog, type *MyToolbar* in the Toolbar Name field, set the Save In value to the current document, and then click on the OK button. This action creates a new, empty toolbar.

3   Select the Commands tab on the Customize dialog.

4   Select Macros from the Categories list to display available macros, as shown in figure 11-10.

Fig. 11-10. Displaying available macros.

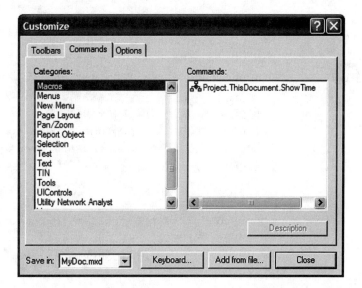

5   Drag the *Project.ThisDocument.ShowTime* command from the Commands list to the toolbar you created in step 2. This action adds the command to the toolbar.

6   Close the Customize dialog and click on the newly created command to see the message box, shown in figure 11-11.

Fig. 11-11. Result of a macro.

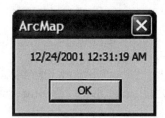

TIP: *You can view the source code for a VBA macro by right-clicking on the custom command while the Customize dialog is open and selecting View Source Code.*

Toolbars

In the previous section you interactively created a new toolbar. The following VBA macro performs the same task of creating a new toolbar and adding the *ShowTime* command.

**CODE
VBA11-1**

```
Public Sub NewToolbar()
  Dim pDocument As IDocument
  Dim pCommandBars As ICommandBars
  Dim pCommandBar As ICommandBar

  Set pDocument = ThisDocument
  Set pCommandBars = pDocument.CommandBars
  Set pCommandBar = pCommandBars.Create("MyToolbar", _
  esriCmdBarTypeToolbar)

  ' Add the ShowTime macro
  pCommandBar.CreateMacroItem "MyCommand", 0, _
  "Project.ThisDocument.ShowTime"
  pCommandBar.Dock esriDockRight
End Sub
```

In the preceding code, the *Create* method of the *ICommandsBar* interface is used to build a new toolbar. This method accepts two arguments: a name for the new toolbar and the bar type. The bar type is identified by one of the following constants.

- *esriCmdBarTypeToolbar* for toolbars
- *esriCmdBarTypeMenu* for menus
- *esriCmdBarTypeShortcutMenu* for context menus

The *ShowTime* macro is added using the *CreateMacroItem* method, which has the following syntax.

```
pCommandBar.CreateMacroItem Name [,FaceID] [,Action] [,Index]
```

Name is the only required argument. It establishes a name for the command. In this example, the name is set to *"MyCommand"*. The optional *FaceID* argument sets the icon for the command. There are 49 predefined images you can use by specifying an index from 0 to 48. The *Action* argument specifies the VBA macro that is executed when the button is pressed. The optional index specifies the location of the command on the toolbar. In this example, the *Action* argument is specified as follows.

```
"Project.ThisDocument.ShowTime".
```

As you can see, the macro name must include the project and module name. The last line of the macro docks the toolbar at the right in the ArcMap window. You can run your macros by selecting Tools | Macros | Macros. This menu displays the Macros dialog, which lists your available macros, as shown in figure 11-12. Highlight the macro you wish to run and click on the Run pushbutton.

Fig. 11-12. Macros dialog box.

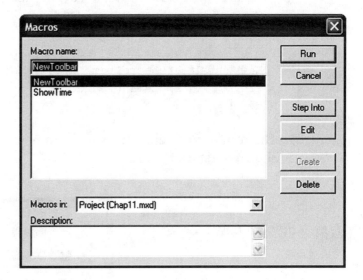

Menus

ArcMap has a Main menu and Standard toolbar that appear by default. The Main menu is a toolbar that contains only menu controls. If you are developing an application with custom pushbutton commands, it is a good practice to make those pushbuttons also available as menus.

You can create a new menu by first creating a new toolbar (as described in the previous section) and then adding menu controls. You can also add new menus to an existing toolbar. The following VBA code shows you how to add a new menu to the Main menu and then add a built-in ArcMap command.

**CODE
VBA11-2**

```
Public Sub NewMenu()
 ' Find the Main menu bar.
 Dim pMainMenuBar As ICommandBar
 Set pMainMenuBar = ThisDocument.CommandBars. _
 Find(ArcID.MainMenu)

 ' Create the new menu called "MyMenu"
 ' on the MainMenuBar.
 Dim pNewMenu As ICommandBar
 Set pNewMenu = pMainMenuBar.CreateMenu("MyMenu")

 ' Add a built in command to the new menu.
 pNewMenu.Add ArcID.File_AddData

 ' Add your VBA macro to the new menu.
 pNewMenu.CreateMacroItem "MyCommand", 0, _
 "Project.ThisDocument.ShowTime"
End Sub
```

The result of running this macro is a new menu in the Main menu, as shown in figure 11-13.

Fig. 11-13. Adding a new menu.

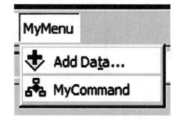

The preceding example uses *ArcID* to find the Main menu and to add the built-in Add Data command. ArcID is a built-in module of the Normal template that returns the unique ID of an ArcMap object. You can view the ArcID code by double clicking on it in the Project Explorer window, shown in figure 11-14.

Fig. 11-14. ArcID module in the Normal template.

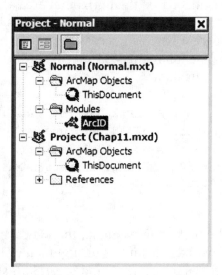

ArcID accepts the name of a command, menu, or toolbar as an argument and returns the UID of that item. The UID is then used by the *Find* or *Add* method of the *ICommandBars* or *ICommandBar* interface. The ArcObjects Developer Help has a long list of objects used by *ArcID*. This list is too long to print here but you can find it under the *Technical Documents/ArcMap: Names and IDs of commands and command bars* help content. Without the ArcID, you must establish the UID from the object's Global Unique ID (GUID), as shown in the following code segment, which finds the Main menu bar.

```
' Find the Main menu bar.
Dim pUID As New UID
Dim pMainMenuBar As ICommandBar
pUID.Value = "{1E739F59-E45F-11D1-9496-080009EEBECB}"
Set pMainMenuBar = ThisDocument.CommandBars. _
Find(pUID)
```

In the preceding code segment, the UID for the Main menu is established by setting the UID's value to the Main menu bar GUID. The GUID values are also listed in the ArcObjects Developer Help.

Status Bar

The status bar is the horizontal area at the bottom of the ArcGIS Application window. You can use the status bar to display messages or show a progress bar. You access the status bar through the *IStatusBar* interface. The following VBA macro shows you how to display a message on the status bar.

CODE
VBA11-3

```
Public Sub ShowStatusBarMessage()
 Dim pStatusBar As IStatusBar
 Set pStatusBar = Application.StatusBar
 pStatusBar.Message(esriStatusMain) = "A message in the _
  status bar"
End Sub
```

In the preceding example, the *Message* method of the *IStatusBar* interface uses a constant as its argument. This constant value is the index of the status bar pane. At index *esriStatusMain*, your message is displayed in the main pane of the status bar. The following are other primary constants for status bar's panes.

• *esriStatusAnimation* for the pane showing the animation icon

• *esriStatusPosition* for the pane showing the mouse position in map coordinates

• *esriStatusPagePosition* for the pane showing the mouse position in page coordinates

Keep in mind that ArcGIS also displays messages in the status bar. For instance, a command's name will be displayed when the mouse pointer is moved over a command. Therefore, you or the user of your macro may not see the message if overwritten by ArcGIS. A good use of messages in the status bar is to show the progress of a long macro by displaying the macro's steps or activities.

You can also show a progress bar in the status bar. A progress bar is useful when running a long macro. The following VBA macro shows you how to display a progress bar in the status bar.

CODE
VBA11-4

```
Public Sub ShowProgressBar()
 ' Show the progress bar
 Dim pStatusBar As IStatusBar
 Set pStatusBar = Application.StatusBar
 pStatusBar.ShowProgressBar "Computing", 0, _
 10000, 1, True
 ' Step through the progress bar
 Dim lngIndex As Long
 For lngIndex = 0 To 10000
  ' Do your computation here
  pStatusBar.StepProgressBar
 Next lngIndex
 MsgBox "Progress bar reached 100%"
 ' Remove progress bar from the status bar
 pStatusBar.HideProgressBar
End Sub
```

The progress bar in the preceding example is set to have 10,000 steps. You set the number of steps based on your macro's activity. For instance, if you were iterating through 725 records of a shape file, the progress bar should have 725 steps. The *ShowProgressBar* method, which displays the progress bar, has the following format.

```
pProgressBar.ShowProgressBar Message, Min, Max, Step, onePanel
```

The message appears to the left of the progress bar on the status bar. The Min and Max values set, respectively, the lowest and highest values. The progress bar always goes from 0% to 100%, the Min value being 0% and the Max value being 100%. Set the Step option's value to indicate the number of values the progress bar will move each time you use the *StepProgressBar* method. Typically, the Step option's value is 1, if you step through the progress bar once at each iteration of your loop. At this time, the *onePanel* argument is not used by the method and must be specified.

Mouse Cursor

In the previous section you learned how to display a progress bar for long processes. Another useful interface customization is to change the shape of the mouse cursor to indicate a change in the state of your program. For example, it is a good idea to change the mouse cursor shape to a "busy" cursor, also known as the hourglass, to indicate that your macro is busy and the user must wait.

You would use the *IMouseCursor* interface to change the mouse cursor shape. This interface has one method, *SetCursor*, which changes the cursor shape. It is a good practice to reset the mouse cursor shape back to the normal shape when appropriate. However, ArcGIS automatically resets the shape when the instance of the interface is released.

The following VBA macro changes the mouse cursor to an hourglass shape for the duration of loop execution. Once the macro is finished, the cursor is automatically reset to the normal shape.

CODE
VBA11-5

```
Public Sub ShowWaitMouseCursor()
  Dim pMouseCursor As IMouseCursor
  Dim lngIndex As Long

  Set pMouseCursor = New MouseCursor
  pMouseCursor.SetCursor 2
  For lngIndex = 0 To 10000
    Application.StatusBar.Message(0) = _
    "Loop counter: " & lngIndex
  Next lngIndex
End Sub
```

The *SetCursor* method in the previous example sets the cursor to a predefined shape. The hourglass is a predefined shape identified by the ID of 2. There are 11 predefined shapes in ArcObjects. The ID of 0 is the normal shape. Figure 11-15 shows the predefined mouse cursor shapes and their IDs.

Fig. 11-15. Predefined mouse cursor shapes.

0 1 2 3

4 5 6 7

8 9 10

You can also set the mouse cursor to a custom shape using a cursor or icon file. Store the custom shape in an *Image* control on a user form and reference it in the *SetCursor* method, as in the following code segment.

```
pMouseCursor.SetCursor UserForm1.Image1.Picture
```

Using Dialog Boxes

When your macro needs to interact with its user to display information or accept input, you can either build user forms or use one of the dialog boxes ArcObjects offers. This section describes the available dialog boxes and how to use them in your macros. Generally it is more efficient to use the dialog boxes than user forms. Figure 11-16 shows a coordinate dialog box.

Fig. 11-16. Coordinate dialog box.

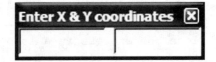

ArcObjects offers the following dialog boxes.

- The Coordinate dialog box (*ICoordinateDialog*) accepts two values from the user.

- The Get String dialog box (*IGetStringDialog*) accepts a text string as input.

- The List dialog box (*IListDialog*) can display a list of text values and return the index of the selected text. Text values are displayed in ascending sort order.

- The Message dialog box (*IMessageDialog*) displays a message.

- The Number dialog box (*INumberDialog*) accepts a numeric value from the user.

- The Password dialog box (*IGetUserAndPasswordDialog*) displays a dialog for user name and password entry.

- The Progress dialog box (*IProgressDialog*) can show an animation and progress bar on a dialog box.

The following code example shows you how to use the Get String and Message dialogs.

CODE
VBA11-6

```
Public Sub UseGetStringDialog()
 Dim pGetStringDialog As IGetStringDialog
 Dim pMessageDialog As IMessageDialog
 Dim blnOK As Boolean
 Dim blnYes As Boolean
 Dim strName As String

 ' Initiate the variables and objects
 strName = "Bob"
 blnYes = False
 Set pGetStringDialog = New GetStringDialog
 Set pMessageDialog = New MessageDialog

 ' Place the process in a loop so that
 ' it can be repeated.
 Do While Not blnYes
  ' Display the Get String Dialog.
  blnOK = pGetStringDialog.DoModal( _
  "Get String Dialog Example", "Name:", strName, 0)
  If blnOK Then
    ' Get the name that user provided.
    strName = pGetStringDialog.Value
    ' Display the Message Dialog.
    blnYes = pMessageDialog.DoModal( _
    "Message Dialog Example", _
    "Is your name " & strName & "?", _
    "Yes", "No", 0)
  Else
    ' User selected the Cancel button,
    ' end the macro.
    Exit Do
  End If
 Loop
End Sub
```

All dialogs, except for the Progress dialog, have a *DoModal* method you can use to display the dialog box. In each instance, the *DoModal* method returns a Boolean value. The returned value indicates the user's action. For example, a false value could mean the user clicked on the Cancel button or typed a bad value. Coordinate or Number dialogs do not have OK or Cancel buttons. In these cases, pressing the Enter key is the same as clicking on the OK button.

All dialogs, except for the Progress and Message dialogs, have one or more properties to get the user value. In the preceding example, the *Value* property of *Get String Dialog* returns the user's input. The Progress dialog is useful when you want to allow canceling of long processes. Figure 11-17 shows the dialog box the following VBA macro creates.

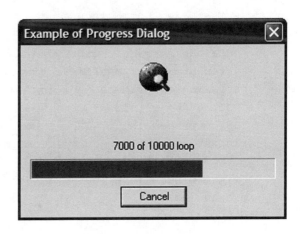

Fig. 11-17. Progress dialog box created by macro code VBA11-7.

CODE VBA11-7

```
Public Sub ShowProgressDialog()
  Dim pProgressDialogFactory As IProgressDialogFactory
  Dim pProgressDialog As IProgressDialog2
  Dim pStepProgressor As IStepProgressor
  Dim pTrackCancel As ITrackCancel
  Dim lngIndex As Long
  Dim blnContinue As Boolean

  ' Use the Progress Dialog Factory to create
  ' a Progress Dialog.
  Set pTrackCancel = New CancelTracker
  Set pProgressDialogFactory = New ProgressDialogFactory
  Set pProgressDialog = pProgressDialogFactory. _
```

```
Create(pTrackCancel, 0)
' Enable the cancel button on the dialog.
pProgressDialog.CancelEnabled = True
pProgressDialog.Title = "Example of Progress Dialog"
' Use a built-in animation.
pProgressDialog.Animation = esriProgressGlobe
blnContinue = True

' Set up the progress bar on the dialog.
Set pStepProgressor = pProgressDialog
pStepProgressor.MinRange = 0
pStepProgressor.MaxRange = 10000
pStepProgressor.StepValue = 1

For lngIndex = 0 To 10000
  ' Carry out the long process here.
  ' Use the Message property to inform
  ' user about the macro's activity.
  pStepProgressor.Message = lngIndex & _
  " of 10000" & " loop"
  ' Stop processing if Cancel button is selected
  blnContinue = pTrackCancel.Continue
  pStepProgressor.Step
  If Not blnContinue Then Exit For
 Next lngIndex
End Sub
```

In the preceding example, a Progress dialog is created from the *IPro-gressDialogFactory* interface. An instance of the *ITrackCancel* interface is passed to the dialog factory so that the Progress dialog can offer a Cancel button. By applying the *Continue* method to the instance of *ItrackCancel*, you can have the macro check if a user has clicked on the Cancel button. This example uses the built-in progress globe. Another built-in animation is identified by *esriDownloadFile*. The Progress dialog is removed from the screen once its instance is released.

CHAPTER **12**

The ArcMap Objects Model

DEVELOPING ARCMAP APPLICATIONS REQUIRES FAMILIARITY with ArcMap's object model. You learned how to read the ArcGIS objects model in Chapter 9. A large portion of that model is dedicated to ArcMap objects. In this chapter you will learn about ArcMap objects used in most applications.

The illustrations in this chapter show only portions of the object model. The entire model is too voluminous to show completely. You can see the entire model by opening the *AllOMDS.PDF* file in the *Object Model Diagrams* directory of the ArcObjects Developer Kit. This chapter will help you understand the ArcMap applications developed in the chapters that follow.

ArcMap Core Objects

The core objects provide access to primary elements of ArcMap, such as maps, layers, and page layout. Figure 12-1 shows the object model for the commonly used core objects.

Fig. 12-1. Commonly used core objects.

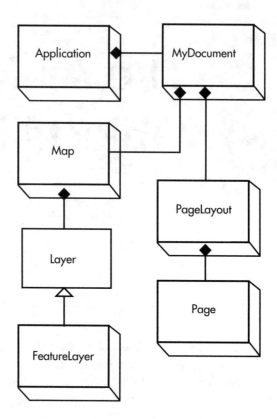

When you start ArcMap, the *Application* object is automatically created. It represents the ArcMap program. In VBA, you access this object using the *Application* keyword. You can instantiate or reference other ArcMap objects from the *Application* object. For example, you can reach a feature layer by first referencing the ArcMap document from the *Application* object. Next, you access the map object from the ArcMap document, and from the map object you can reach the feature layer.

MxDocument is the ArcMap document. This object is also created when you start ArcMap. The ArcMap document creates and manages data representation objects such as the *Map* and *PageLayout* objects. ArcMap's view is either the map or the layout view. The map view corresponds to the *Map* object, and the layout view corresponds to the *PageLayout* object. The *ActiveView* property in the *IMxDocument* interface determines which view is displayed.

Each ArcMap document contains one or more *Map* objects. A *Map* object manages layers that display maps or images in the ArcMap view. Only one map in ArcMap can be activated. The activated map is known as the focus map. You can use the *Layer* or *Layers* property of the *IMap* interface to reference one of the map's layers.

There are several types of layers, such as feature, image, and annotation layers. *Layer* is an abstract class and cannot be instantiated. Its purpose is to specialize into creatable classes of layer types. The *FeatureLayer* object represents spatial data in a vector-based geographic data set such as a shape file, coverage, or geodatabase.

The *PageLayout* object can be referenced through the *PageLayout* property of the *IMxDocument* interface. The *PageLayout* object has a *Page* object that models the hard-copy output page.

ArcMap Data Window Objects

ArcMap's data windows display additional data, such as an overview map. Figure 12-2 shows the object model for the Overview and Map Inset data windows.

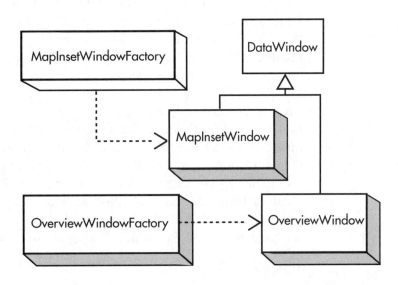

Fig. 12-2. Overview and Map Inset data windows objects model.

The *MapInsetWindow* object is the magnifier window that provides a zoomed view of the activated map. Interactively, you can open the magnifier window by selecting Windows | Magnifier. *MapInsetWindow* is not a co-class and therefore cannot be created directly. You must use *MapInsetWindowFactory* to create an object of *MapInsetWindow*.

The *OverviewWindow* object is the same as the overview window you open by selecting Windows | Overview. It shows the full extent of the activated map, with a box representing the view's extent. *Overview-Window* is not a co-class and therefore cannot be created directly. You must use *OverviewWindowFactory* to create an object of *OverviewWindow*.

ArcMap Element Objects

Elements in ArcMap appear in the layout view. They can be graphic elements, such as a rectangle, or frame elements, such as map frame. The graphic elements are non feature-based shapes that appear on a map or layout. Frame elements provide the border and background for containing other map elements. Figure 12-3 shows a portion of the Arc-Map elements object model.

You can use objects of *TextElement* to place text in the layout view. The *ITextElement* interface provides for setting the text value and its symbol. The *FillShapeElement* class specializes into several 2D, closed-area graphic shapes, such as a polygon. You can set the geometry and symbology of these shapes.

ArcMap displays only the activated *Map* object in the map view. However, it displays all *Map* objects in the layout view. It displays each map inside a *MapFrame* object. *MapSurroundFrame* stores and displays map-surround objects such as north arrows and scale bars. A *MapSurroundFrame* object can be associated with a *MapFrame* object so that changes in the *MapFrame* object will be automatically reflected. For example, the value displayed by a scale-bar surround should change as the map in a *MapFrame* object is zoomed in or out.

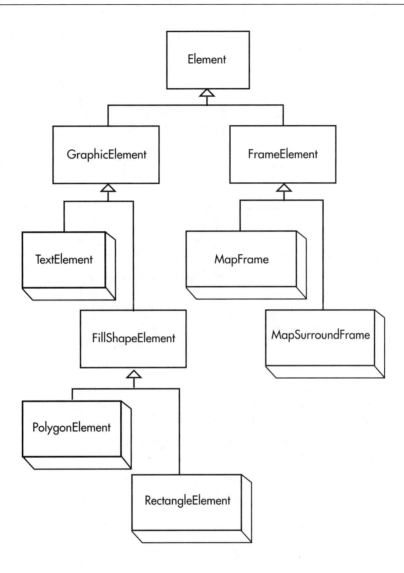

Fig. 12-3. Partial model of the element objects.

ArcMap Map-surround Objects

Map surrounds are elements associated with a map, such as north arrow and scale bar. You can place map surrounds on your layout view by storing them in *MapSurroundFrames* and adding the frames to the view. Figure 12-4 shows the primary objects in the map-surround object model.

Fig. 12-4. Primary objects of the map-surround object model.

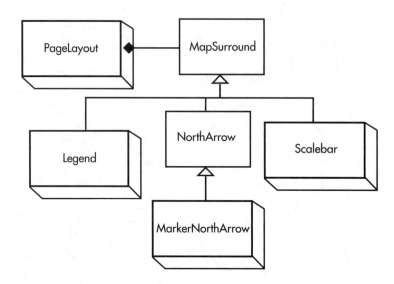

Objects of the *Legend* co-class display a map's legend. The *MarkerNorthArrow* co-class offers north arrow map surrounds based on marker symbols. The *ScaleBar* co-class provides for creating a variety of scale bars.

CHAPTER **13**

Using Maps

OBJECTS OF THE MAP CO-CLASS CONTAIN THE MAP DATA needed to display and manipulate maps. In an ArcMap document you can have one or more *Map* objects. In the data view of ArcMap, only the activated *Map* object is displayed. The activated *Map* object is also known as the focus map. In the layout of ArcMap, all *Map* objects are displayed in their own frame. In addition to the feature or image layers a map can have, each *Map* object also contains a graphic layer. You can see that the *Map* object is one of the primary objects needed in every ArcGIS application that deals with maps.

In this chapter you will learn how to access and use the *Map* object of ArcMap. This chapter's code examples will show you how to select features by location, add graphics, and change the extent of the map display. Collectively, the code examples constitute a simplified application for permits.

Defining the Application

Local governments are responsible for managing and enforcing zoning regulations. Often when a special-use permit is issued or an owner requests a zoning change, owners of the nearby properties must be notified of the change. A simple GIS application that would select such nearby properties would facilitate the process. In this chapter you will develop an application having the following capabilities.

- The user clicks on the property containing the request to change its zoning and provides a buffer distance. The zoning laws generally define the buffer distance. Owners of the properties that fully or partially fall within this buffer distance must be notified.

- The application creates the buffer around the identified property and adds the buffer limit to the map as a graphic object.

- The application then selects the properties that are within or intersect the buffer; and finally the application changes the map extent to the extent of the selected properties.

You need to enter the application as a VBA macro into an ArcMap document. This ArcMap document must have an activated map with a single layer of properties named *Parcels*. ArcGIS installation comes with tutorial data that has a *Parcels* layer. You can use that or your own *Parcels* layer.

Accessing the Map Object

New *Map* objects can be created, or existing *Map* objects can be accessed through the *IMxDocument* interface. The following code segment shows you how to create a new *Map* object.

CODE
VBA13-1

```
Dim pMxDocument As IMxDocument
Dim pMaps As IMaps
Dim pMap As IMap
Set pMxDocument = Application.Document
Set pMaps = pMxDocument.Maps
Set pMap = New Map
pMap.Name = "My New Map"
pMaps.Add pMap
```

In the preceding example, a new *Map* object is created and added to the Maps collection of the ArcMap document. When your application expects to work with an existing map, you can access the *Map* object using the following code segment.

CODE
VBA13-2

```
Dim pMxDocument As IMxDocument
Dim pMap As IMap
Set pMxDocument = Application.Document
Set pMap = pMxDocument.FocusMap
```

In the preceding example, the *FocusMap* property of the ArcMap document returns the activated map.

Selecting by Location

The first task in our development process is to create the user interface. A tool button would be appropriate for this application because you need the user to click on the property with the permit request. When the user selects the tool button and clicks on a feature of the *Parcels* layer, the application selects the feature and starts the process of creating the buffer.

ArcMap has a tool for selecting features. You could develop the application such that the user would have to first select a land parcel. However, in this application you will write your own selection tool so that you can continue with the rest of the process once the user has identified the parcel.

Start by creating a new toolbar. Chapter 11 showed you how to add a new toolbar from the Customize dialog. The following steps take you through the process of adding a tool button to your new toolbar.

1   In the Customize dialog, select the Commands tab. Then select UIControls in the Categories list. Select your current document in the *Save in* combo box.

2   Click on the New UIControl button to open the New UIControl dialog. Select the UIToolControl option and click on the Create button. ArcMap closes the UIControl dialog and places *Project.UIToolControl1* in the Commands list of the Customize dialog.

3   Rename the *Project.UIToolControl1* object by first selecting it in the Commands list and then clicking on it again. Do not double click. Change the name to *Project.UIToolPermit*, as shown in figure 13-1.

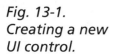

Fig. 13-1.
Creating a new
UI control.

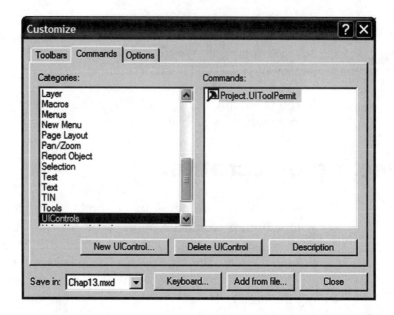

4 Drag *Project.UIToolPermit* from the Customize dialog into your new toolbar. Close the Customize dialog.

You now have a new tool button for the application's interface. You now need to program the application that executes when the user clicks on the map with the new tool button selected. Add the programming pieces as follows.

1 Right-click on the new tool button to display its context menu. Select View Source. This action opens the VBA editor and places you in the *UIToolPermit_Select()* procedure. ArcMap has automatically created this procedure. This procedure is executed when the tool is selected. You can create other procedures for other events, as in step 2.

2 The *UIToolPermit_Select()* is not needed, so you can delete or just ignore it. Instead, you need procedures for when the mouse is clicked on the map. On top of the Code window there are two combo boxes. The left-hand combo box shows objects and should be set at *UIToolPermit*. The right-hand combo box lists procedures. For the right-hand combo box, select the *MouseDown* procedure to create *UIToolPermit_MouseDown*. Then select the *MouseUp* procedure from the right-hand combo box to create the

UIToolPermit_MouseUp. These two procedures, as their names imply, execute when the user clicks the mouse button while the application tool is selected.

3 Add the following code lines to the top of the Code window above the first procedure.

**CODE
VBA13-3**

```
Option Explicit
Dim m_blnMouseDown As Boolean
Dim m_pPoint As IPoint
```

The *Option Explicit* statement forces you to declare all variables. It reduces errors caused by mistyping. The two decaled variables will be used by the procedures of the application. Because these two variables are outside the procedures, their scope is at the application level. This means that while the application is running any procedure can access them.

4 Add the programming code to the *MouseDown* procedure, as shown in the following code listing.

**CODE
VBA13-4**

```
Private Sub UIToolPermit_MouseDown(ByVal button As Long, _
ByVal shift As Long, ByVal x As Long, ByVal y As Long)
    ' The process of selction begins in this
    ' procedure by pressing the mouse button down
    ' on the map. It ends in the MouseUp event procedure.
    Dim pMxDocument As IMxDocument
    Dim pActiveView As IActiveView
    m_blnMouseDown = True
    Set pMxDocument = Application.Document
    Set pActiveView = pMxDocument.FocusMap
    Set m_pPoint = pActiveView.ScreenDisplay. _
    DisplayTransformation.ToMapPoint(x, y)
End Sub
```

In the preceding procedure, the *m_blnMouseDown* variable is used to track where the user clicked down on the mouse button. If the user clicked outside the map, *m_blnMouseDown* is not set and the *MouseUp* procedure does not act on the map. ArcMap passes to the procedure the location of the mouse click in the *x* and *y* parameters. The last statement converts the location from display coordinates to map coordinates.

5   Add the programming code to the *MouseUp* procedure, as shown in the following code listing.

CODE VBA13-5

```
Private Sub UIToolPermit_MouseUp(ByVal button As Long, _
ByVal shift As Long, ByVal x As Long, ByVal y As Long)
  ' Do not process if the mouse was not pressed down
  ' on the map.
  If (Not m_blnMouseDown) Then Exit Sub
  Dim pMxDocument As IMxDocument
  Dim pActiveView As IActiveView
  Dim pMap As IMap
  Set pMxDocument = Application.Document
  Set pActiveView = pMxDocument.FocusMap
  pActiveView.PartialRefresh esriViewGeoSelection, _
  Nothing, Nothing
  Set pMap = pMxDocument.FocusMap
  pMap.SelectByShape m_pPoint, Nothing, True
  pActiveView.PartialRefresh esriViewGeoSelection, _
  Nothing, Nothing
  m_blnMouseDown = False
End Sub
```

First, the *MouseUp* procedure checks the *m_blnMouseDown* variable. If it is not set, the user did not press the mouse button on the map. This variable is reset at the end of the procedure. The *MouseDown* procedure stored the location of the mouse click in the *m_pPoint* object. This procedure uses that object in the *SelectByShape* method of *Map* to select a feature. The procedure also partially refreshes the active view twice. The first refresh clears any old selection, and the next refresh updates the view with the new selection.

The *PartialRefresh* method needs three parameters: phase, data, and envelope. In the preceding procedure, the phase of the partial refresh is set to *esriViewGeoSelection*. The partial refresh is faster than complete refresh because it only refreshes the specified phase. In this case, the selected geography is refreshed.

6   You are now ready to test this part of the application. Save your work and from the ArcMap window select the Permit Application tool button. Then click on a feature of the *Parcels* layer. The parcel will be selected.

You need to add one more procedure related to selection by location. In the next section, you will see how the buffer polygon is created. The application uses the buffer polygon to select new parcels. The following procedure accepts the buffer polygon and selects the intersecting parcels. Add this procedure to the end of the Code window in the VBA editor.

CODE
VBA13-6

```
Private Sub SelectParcels(pBufferGeometry As IGeometry)
  ' Select the parcels that intersect the buffer
  ' except for the original parcel.
  Dim pMxDocument As IMxDocument
  Dim pActiveView As IActiveView
  Dim pMap As IMap
  Set pMxDocument = Application.Document
  Set pActiveView = pMxDocument.FocusMap
  pActiveView.PartialRefresh esriViewGeoSelection, _
  Nothing, Nothing
  Set pMap = pMxDocument.FocusMap

  Dim pSelectionEnvironment As ISelectionEnvironment
  Set pSelectionEnvironment = New SelectionEnvironment
  pSelectionEnvironment.AreaSelectionMethod = _
  esriSpatialRelIntersects
  pSelectionEnvironment.CombinationMethod = _
  esriSelectionResultXOR
  pMap.SelectByShape pBufferGeometry, _
  pSelectionEnvironment, False
  pActiveView.PartialRefresh esriViewGeoSelection, _
  Nothing, Nothing
End Sub
```

The *SelectByShape* method of the *Map* object is again used in this procedure. However, this time a selection environment is defined as the second parameter of the method. When no selection environment is given, the map's default is used. Because you cannot be certain what the map's selection environment is at the time of execution, you need to define one.

This selection environment sets the area selection method and selection combination method. The application uses intersection as the area selection method. It sets the combination method to XOR. This combination method selects if a feature is not selected, and deselects if a feature is selected. By using XOR, you select the nearby parcels and deselect the original parcel.

Adding Graphics

Graphic elements are added to the graphic layer of the map. This layer is stored with the map document. The graphic layer holds labels and graphic shapes. In the Permit application, you build the buffer and add it to the map as a polygon graphic. In the VBA editor, add the following procedure to the end of the Code window.

CODE
VBA13-7

```
Private Function BufferFeatures() As IGeometry
    Dim pMxDocument As IMxDocument
    Dim pActiveView As IActiveView
    Dim pMap As IMap
    Dim GraphicsContainer As IGraphicsContainer
    Dim strDistanceUnit As String
    Dim strBufferDistance As String

    Set pMxDocument = Application.Document
    Set pActiveView = pMxDocument.FocusMap
    Set pGraphicsContainer = pMxDocument.FocusMap

    ' Verify a feature is selected
    Set pMap = pMxDocument.FocusMap
    If pMap.SelectionCount = 0 Then Exit Function

    ' Get the buffer distance
    strBufferDistance = _
    InputBox("Enter Buffer Distance:", "")
    If IsNull(strBufferDistance) Or _
    Not IsNumeric(strBufferDistance) Then Exit Function
```

```
' Build the symbol for the buffer
Dim pElement As IElement
Dim pFillShapeElement As IFillShapeElement
Dim pFillSymbol As IFillSymbol
Dim pColor As IColor
Dim pLineSymbol As ILineSymbol
Set pElement = New PolygonElement
Set pFillShapeElement = pElement
Set pFillSymbol = pFillShapeElement.Symbol
Set pColor = pFillSymbol.Color
Set pLineSymbol = pFillSymbol.Outline
pColor.Transparency = 0
pFillSymbol.Color = pColor
pColor.Transparency = 255
pColor.RGB = RGB(255, 0, 0)
pLineSymbol.Color = pColor
pLineSymbol.Width = 0.1
pFillSymbol.Outline = pLineSymbol
pFillShapeElement.Symbol = pFillSymbol


' Buffer the first selected feature
Dim pEnumFeature As IEnumFeature
Dim pTopoOperator As ITopologicalOperator
Dim pFeature As IFeature
Set pEnumFeature = _
pMxDocument.FocusMap.FeatureSelection
pEnumFeature.Reset
Set pFeature = pEnumFeature.Next
Set pTopoOperator = pFeature.Shape
pElement.Geometry = pTopoOperator. _
Buffer(CInt(strBufferDistance))
pGraphicsContainer.AddElement pElement, 0
pActiveView.PartialRefresh esriViewGraphics, _
Nothing, Nothing

' Return the buffer's geometry
Set BufferFeatures = pElement.Geometry
End Function
```

The preceding procedure is defined as a function that returns the buffer as a *Geometry* object. Later, this object is used to select features from the *Parcels* layer. The *BufferFeatures* function first checks to make sure a feature is selected. You can check for the number of selected features by examining the *SelectionCount* property of the Map object. Next, the function asks the user to enter a buffer distance.

The permit application should show the buffer polygon as a transparent shape with a red outline. In this fashion, the buffer does not obscure the parcels it covers. The next step in the function is setting the symbol for the buffer polygon. When the buffer is created, you will assign it to a graphic element object. Before calculating the buffer, you declare the element as an *IElement* and set its symbology.

The *IFillShapeElement* interface provides access to the fill and outline *Symbol* objects of a graphic element. Once the *Symbol* objects are available, you can set their color and other properties. You need to reset the symbol property of the graphic elements after you have changed the colors.

Next, you calculate the buffer and assign the resulting polygon to the graphic element you created earlier. The *FeatureSelection* method of the *Map* object returns a collection of selected features. In ArcMap you work with collections through enumerators. An enumerator provides methods for iterating through the collection and accessing collection members.

The *FeatureSelection* method returns a reference to an *IEnumFeature* interface. This interface, similar to other enumerators, has two methods: *Next* and *Reset*. Once you have the enumerator, you must first use the *Reset* method to point to the first element. Then, use the *Next* method to sequentially retrieve the enumerator's members. In this application, for simplicity, the first selected feature is used.

The methods in the *ITopologicalOperator* interface let you create new geometries based on topological relationships between existing geometries. Once the buffer geometry is available through the graphic element, it is added to the graphic layer of the map using the *AddElement* object of the *IGraphicsContainer* interface.

Changing the Extent

GIS applications frequently need to change the map extent to display a specific portion of the map. In the Permit application you want to change the extent so that the selected nearby parcels are shown in the map. The application changes the map extent property by setting it to a new *Envelope* object. In the next procedure, the extents for all selected parcels are merged into a single *Envelope* object. The resulting *Envelope* object is then used to reset the map's extent. Add the following procedure to the Code window in the VBA editor.

CODE VBA13-8

```vba
Private Sub SetExtent()
  Dim pMxDocument As IMxDocument
  Dim pActiveView As IActiveView
  Dim pEnumFeature As IEnumFeature
  Dim pFeature As IFeature
  Dim pElement As IElement
  Dim pEnvelope As IEnvelope

  Set pMxDocument = Application.Document
  Set pActiveView = pMxDocument.FocusMap

  If pMxDocument.FocusMap.SelectionCount = 0 _
  Then End

  Set pEnvelope = New Envelope
  Set pEnumFeature = pMxDocument.FocusMap. _
  FeatureSelection
  pEnumFeature.Reset
  Set pFeature = pEnumFeature.Next
  Do While Not pFeature Is Nothing
    pEnvelope.Union pFeature.Extent
    Set pFeature = pEnumFeature.Next
  Loop

  pActiveView.Extent = pEnvelope
  pActiveView.Refresh

End Sub
```

In the preceding procedure, you loop through the enumeration to retrieve the selected features and merge their envelopes. The property is then reset to the *Envelope* object, which encompasses all selected features.

Testing the Application

You need to add a few more line of code before you can test the application. The application now has the following procedures.

- *UIToolPermit_MouseDown*
- *UIToolPermit_MouseUp*
- *SelectParcels*
- *BufferFeatures*
- *SetExtent*

The application starts with the *UIToolPermit_MouseDown* procedure and continues with *UIToolPermit_MouseUp* to select the parcel the user clicked on. However, the application should continue with *BufferFeatures* to create the buffer, *SelectParcels* to select nearby parcels, and *SetExtent* to change the map extent. For these procedures to execute, you need to call them from the *UIToolPermit_MouseUp* procedure. Add the following code segment to the end of the *UIToolPermit_MouseUp* procedure.

**CODE
VBA13-9**

```
' After a parcel is selected, create a buffer around it
Dim pBufferGeometry As IGeometry
Set pBufferGeometry = BufferFeatures()
' Use the buffer to select parcels for notification
SelectParcels pBufferGeometry
' Zoom to the selection
SetExtent
```

Make sure the *Parcels* layer is selectable. You can check if a layer is selectable by selecting Selection | Set Selectable Layers. Clear all selected features, save your work, and try the application. Figure 13-2 shows the result of the application.

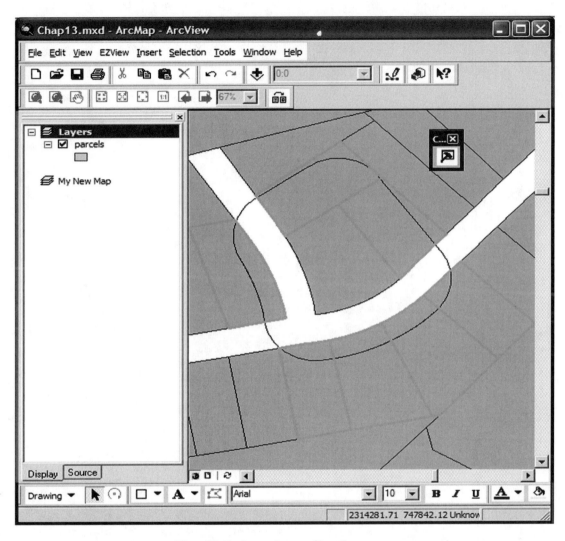

Fig. 13-2. Permit application.

14

Using Layers

AN ARCMAP DOCUMENT CAN CONTAIN one or more maps. Each map may have one or more layers. ArcMap displays geographic data through layers. A layer does not store the data; rather, it references the geographic data stored outside ArcMap. ArcMap can display a variety of geographic data, such as shape files, geodatabases, images, CAD data, and coverages.

Your ArcMap application can use the *Layer* object to query or label features. This chapter shows you how to access the *Layer* object, load a shape file, perform an attribute query, and label a layer. Collectively, the code examples are for a simplified application described in the material that follows.

Defining the Application

A common GIS functionality that many organizations provide to their users is to navigate to a location on the map based on a name. For example, a GIS application in the planning office may have a menu item for zooming to a specific planning district. In this chapter, you will develop an application that finds and zooms to an airport. The application has the following functionality.

- Access the Airports layer or load it.

- Get an airport name from the user and perform a query.

- Zoom to the selected airport and label it.

The CD-ROM number 3 of the ESRI data distributed with ArcGIS has the airport shape file you can use. You enter this application into Arc-Map as a VBA macro. You then run the application by selecting Tools | Macros | Macros.

Start by opening the VBA editor, selecting *ThisDocument* from the Project window, and selecting View | Code to display the Code window for *ThisDocument*. In the empty Code window, enter the following statement.

```
Option Explicit
```

The preceding statement forces you to declare all variables. This reduces the problems caused by mistyping variables.

Finding a Layer

To find a layer you must loop through the existing layers searching for the desired one. You can search by the layer's name, but if the user of your application changes the names of layers, your application may fail. You may want to use the *BrowseName* property when searching for a layer that has a shape file source. This property returns the file name, which does not change even if the layer name has changed. Add the following procedure to the Code window of *ThisDocument*.

CODE VBA14-1

```
Private Function FindLayer(strBrowseName As String) _
As ILayer
    ' Find the matching layer in the focus map.
    ' Return NOTHING if not found.
    Dim lngIndex As Long
    Dim pMxDocument As IMxDocument
    Dim pMap As IMap
    Dim pDataset As IDataset
    Set FindLayer = Nothing
    Set pMxDocument = Application.Document
    Set pMap = pMxDocument.FocusMap
    ' Loop through layers and search
    ' for the matching name.
```

```
    For lngIndex = 0 To (pMap.LayerCount - 1)
      If TypeOf pMap.Layer(lngIndex) Is FeatureLayer Then
        Set pDataset = pMap.Layer(lngIndex)
        If pDataset.BrowseName = strBrowseName Then
          Set FindLayer = pMap.Layer(lngIndex)
          Exit For
        End If
      End If
    Next lngIndex
End Function
```

The preceding procedure is a function that searches the focus map for a layer that matches the given *BrowseName.* The function returns the layer if it finds it. Otherwise, it returns *Nothing.* In this example, the search is limited to the focus map. Note that you could loop through all maps and search the layers of each map if you did not want to limit your search to the activated map. The loop for iterating through the maps might take the following form.

```
Dim pMxDocument As IMxDocument
Dim pMaps As IMaps
Dim pMap As IMap
Dim lngLoop As Long
Set pMxDocument = Application.Document
Set pMaps = pMxDocument.Maps
For lngLoop = 0 To pMaps.Count - 1
  Set pMap = pMaps.Item(lngLoop)
Next lngLoop
```

Loading a Shape File

As discussed in the previous section, a user can change a layer's name. A user can also remove a layer your application may depend on. Therefore, you must always check for missing critical layers, and load them if they are not available. Add the following procedure to the Code window. This procedure loads a shape file based on the given path and file name.

**CODE
VBA14-2**

```
Private Function LoadShapefile(strPath As String, _
strFile As String, strName As String) As ILayer
   ' Load the shapefile based on the given path and name.
   ' Shapefile is added to the focus map as a layer.
   ' Layer is returned when successful.
   Dim pWorkspaceFactory As IWorkspaceFactory
   Dim pFeatureWorkspace As IFeatureWorkspace
   Dim pFeatureClass As IFeatureClass
   Dim pFeatureLayer As IFeatureLayer
   Dim pMxDocument As IMxDocument
   Dim pMap As IMap
   Set pWorkspaceFactory = New ShapefileWorkspaceFactory
   Set pFeatureWorkspace = pWorkspaceFactory. _
   OpenFromFile(strPath, 0)
   Set pFeatureClass = pFeatureWorkspace. _
   OpenFeatureClass(strFile)
   Set pFeatureLayer = New FeatureLayer
   Set pFeatureLayer.FeatureClass = pFeatureClass
   ' Set the name and visibility of the layer.
   pFeatureLayer.Name = strName
   pFeatureLayer.Visible = True
   ' Add the shapefile to the focus map.
   Set pMxDocument = Application.Document
   Set pMap = pMxDocument.FocusMap
   pMap.AddLayer pFeatureLayer
   Set LoadShapefile = pFeatureLayer
End Function
```

In the preceding function, the *Workspace* object is used to open the shape file. There are two types of workspaces: file system and database. The file system workspace manages data sets such as shape files and coverages. Database workspaces maintain geodatabases. *Workspace* is not a creatable class, and can be instantiated only through objects of *WorkspaceFactory*. There are many types of *WorkspaceFactory*. *ShapefileWorkspaceFactory* is used here. The following code segment shows you how to add a feature class from a personal geodatabase.

```
Dim pWorkspaceFactory As IWorkspaceFactory
Dim pFeatureWorkspace As IFeatureWorkspace
Dim pFeatureClass As IFeatureClass
```

```
Dim pFeatureLayer As IFeatureLayer
Dim pMxDocument As IMxDocument
Dim pMap As IMap
Set pWorkspaceFactory = New AccessWorkspaceFactory
Set pFeatureWorkspace = pWorkspaceFactory. _
OpenFromFile("C:\myGDB.mdb", 0)
Set pFeatureClass = pFeatureWorkspace. _
OpenFeatureClass("myFC")
Set pFeatureLayer = New FeatureLayer
Set pFeatureLayer.FeatureClass = pFeatureClass
' Set the name and visibility of the layer.
pFeatureLayer.Name = "My Feature Class"
pFeatureLayer.Visible = True
' Add the shapefile to the focus map.
Set pMxDocument = Application.Document
Set pMap = pMxDocument.FocusMap
pMap.AddLayer pFeatureLayer
```

Querying a Layer

You can use the objects of the *QueryFilter* co-class to retrieve a subset of tabular data. This application uses *QueryFilter* to select an airport by its name. Add the following procedure to the Code window.

CODE VBA14-3

```
Private Sub FindFeature(pFeatureLayer As IFeatureLayer, _
strSearch As String)
  ' Select features based on the given search value.
  Dim pMxDocument As IMxDocument
  Dim pMap As IMap
  Dim pActiveView As IActiveView
  Dim pFeatureSelection As IFeatureSelection
  Dim pQueryFilter As IQueryFilter
  Set pMxDocument = Application.Document
  Set pMap = pMxDocument.FocusMap
  Set pActiveView = pMap
  Set pFeatureSelection = pFeatureLayer
  'Create the query filter
  Set pQueryFilter = New QueryFilter
  pQueryFilter.WhereClause = "NAME like '%" & _
  strSearch & "%'"
```

```
' Clear previous selection.
pActiveView.PartialRefresh esriViewGeoSelection, _
Nothing, Nothing
'Do the query and refresh the map.
pFeatureSelection.SelectFeatures pQueryFilter, _
esriSelectionResultNew, False
pActiveView.PartialRefresh esriViewGeoSelection, _
Nothing, Nothing
End Sub
```

The *IFeatureSelection* interface provides the *SelectFeatures* method, which can query a layer. This is an interface of *FeatureLayer*. Therefore, to reference it you query the *IFeatureLayer* interface, as shown in the following code segment.

```
Set pFeatureSelection = pFeatureLayer
```

The "like" operator is used in the *where* clause of the query filter. In this manner, the user does not have to type the entire airport name. For example, if the user enters *Dulles*, the application will find *Washington Dulles International*.

Zooming to Selected Features

To set the extent of the map to display the selected features, the extents of all selected features are combined. The application uses an enumerator to iterate through the selected features. Add the following procedure to the Code window.

CODE
VBA14-4

```
Private Function ZoomToSelected() As Boolean
    ' Zoom to the selection.
    Dim pMxDocument As IMxDocument
    Dim pMap As IMap
    Dim pActiveView As IActiveView
    Set pMxDocument = Application.Document
    Set pMap = pMxDocument.FocusMap
    ZoomToSelected = False
    ' Do not change the extent if there are
    ' no selected features.
    If pMap.SelectionCount > 0 Then
```

```
      Dim pEnumFeature As IEnumFeature
      Dim pFeature As IFeature
      Dim pEnvelope As IEnvelope
      ' Retrieve the selected features and
      ' get selections' extent.
      Set pEnumFeature = pMap.FeatureSelection
      pEnumFeature.Reset
      Set pFeature = pEnumFeature.Next
      Set pEnvelope = New Envelope
      Do While Not pFeature Is Nothing
         ' The new extent is combination
         ' of selected feature's extents.
         pEnvelope.Union pFeature.Extent
         Set pFeature = pEnumFeature.Next
      Loop
      pEnvelope.Expand 1.1, 1.1, True
      ' Set map's extent to the
      ' selections' extent.
      Set pActiveView = pMap
      pActiveView.Extent = pEnvelope
      pActiveView.Refresh
      ZoomToSelected = True
   End If
End Function
```

You access an enumeration through a method of an object. In the preceding procedure, the *FeatureSelection* method of the *IMap* interface returns a reference to the *IEnumFeature* enumerator. Enumerators have two methods: *Reset* and *Next*. Always use the *Reset* method first, to set the enumeration to its beginning. Then use the *Next* method to retrieve the next object from the enumerator. When there are no more objects, *Next* returns *Nothing*.

The application uses instances of the Envelope co-class to manage extents. The *Union* method of the *IEnvelope* interface combines the extent of the selected features into a new envelope. The combined envelope is then used to set the map extent.

Labeling a Layer

The easiest way to create labels is to turn on the labeling property of a layer. The *IGeoFeatureLayer* interface has the methods for setting the label field and turning the labeling on or off. Add the following procedure to the Code window of *ThisDocument*.

CODE VBA14-5

```
Private Sub LabelLayer(pFeatureLayer As IFeatureLayer, _
blnLabel As Boolean)
    ' Label a layer by turning on
    ' the label property.
    Dim pGeoFeatureLayer As IGeoFeatureLayer
    Dim pMxDocument As IMxDocument
    Dim pMap As IMap
    Dim pActiveView As IActiveView
    Set pMxDocument = Application.Document
    Set pMap = pMxDocument.FocusMap
    Set pActiveView = pMap
    ' Turn on or off the layer's labeling.
    Set pGeoFeatureLayer = pFeatureLayer
    pGeoFeatureLayer.DisplayField = "NAME"
    pGeoFeatureLayer.DisplayAnnotation = blnLabel
    pActiveView.Refresh
End Sub
```

In the preceding procedure, once the labeling property is turned on or off by the *DisplayAnnotation* method, the view is refreshed.

Testing the Application

So far in this chapter you have been creating individual procedures that perform various parts of the application. You need one more procedure to bring together these parts. This last procedure should be a public sub so that it could be executed as a macro.

CODE VBA14-6

```
Public Sub LocateAirport()

    ' (1) Get the Airports layer.
    Dim pFeatureLayer As IFeatureLayer
```

```
Set pFeatureLayer = FindLayer("airports")
If pFeatureLayer Is Nothing Then
   Set pFeatureLayer = LoadShapefile _
   ("c:\arcgis_data\cd3\usa", "airports", _
   "Airport")
End If

' (2) Ask user for the airport name.
Dim pGetStringDialog As IGetStringDialog
Set pGetStringDialog = New GetStringDialog
If Not pGetStringDialog.DoModal _
("", "Airport:", "", 0) Then
   Exit Sub
End If

' (3) Select the airport
Dim strName As String
strName = UCase(pGetStringDialog.Value)
If strName > "" Then
   FindFeature pFeatureLayer, strName
End If

' (4) Zoom to the selected features
'    and label the airport.
If ZoomToSelected() Then
   LabelLayer pFeatureLayer, True
Else
   LabelLayer pFeatureLayer, False
End If

End Sub
```

The preceding procedure begins by getting the airports layer. If the layer does not exist, it loads it from its shape file. In the second step, the application uses a dialog box to get the airport name from the user. The third step performs a query to select the airport. Finally, the view's extent is changed to the extent of the selected airport and labels are turned on.

Save your work and run the macro by selecting Tools | Macros | Macros from the ArcMap window. Once the Macros dialog is shown, select *LocateAirport* and click on the Run button to start the application.

CHAPTER **15**

Using Data Windows

ARCMAP OFFERS TWO ADDITIONAL WINDOWS that can display spatial data: Overview and Magnification. The Overview window shows the full extent of your map, with a box representing the current display. The Magnification window acts like a magnifier. As you move the window over your map, you can see part of your map at a different scale. Figure 15-1 shows both windows. You can open these windows in ArcMap by selecting the Window menu option. You can open multiple Overview or Magnification windows.

In your application you can use the Overview window to allow the user to quickly pan to areas of the map that are not in the display. You can use the Magnification window to show details without zooming in. If a layer is not shown because of the map scale, the magnification window can show it so that you do not have to change the map scale. This chapter shows you how to create and use these two windows.

Accessing Data Windows

Your application can create new data windows, or it may need to access the data windows already open. The *IApplicationWindows* interface of the *Application* object provides the *DataWindows* method for accessing opened data windows. This method returns a collection in the form of a set. You can use the *Reset* and *Next* methods of the *ISet* interface to iterate through the data windows. The following VBA macro shows you how to access the set of opened data windows.

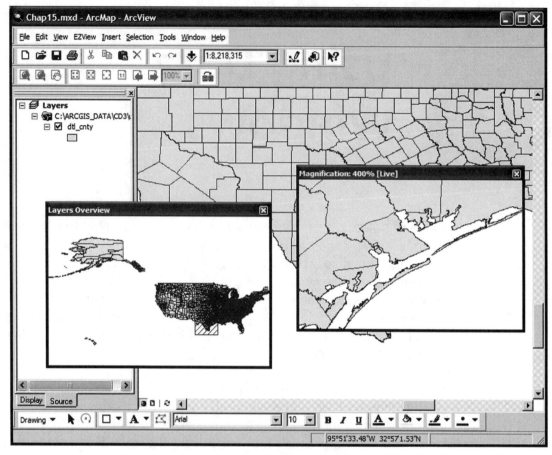

Fig. 15-1. Overview and Magnification windows.

**CODE
VBA15-1**

```
Public Sub TileDataWindows()
    Dim pApplicationWindows As IApplicationWindows
    Dim pWindowsSet As ISet
    Dim pDataWindow As IDataWindow
    Dim lngLeft As Long
    Dim lngTop As Long
    Dim lngRight As Long
    Dim lngBottom As Long
    Dim lngNewLeft As Long
    Dim lngNewTop As Long
    Dim lngNewRight As Long
    Dim lngNewBottom As Long
```

```
    Dim lngWidth As Long
    Dim lngHeight As Long

    lngNewLeft = 0
    lngNewTop = 0
    Set pApplicationWindows = Application
    Set pWindowsSet = pApplicationWindows.DataWindows
    pWindowsSet.Reset
    Set pDataWindow = pWindowsSet.Next
    Do While Not pDataWindow Is Nothing
      pDataWindow.QueryPosition lngLeft, lngTop, _
      lngRight, lngBottom
      lngWidth = lngRight - lngLeft
      lngHeight = lngBottom - lngTop
      lngNewRight = lngNewLeft + lngWidth
      lngNewBottom = lngNewTop + lngHeight
      pDataWindow.PutPosition lngNewLeft, lngNewTop, _
      lngNewRight, lngNewBottom
      lngNewLeft = lngNewRight + 1
      If TypeOf pDataWindow Is IMapInsetWindow Then
        pDataWindow.Refresh
      End If
      Set pDataWindow = pWindowsSet.Next
    Loop
End Sub
```

The preceding macro arranges the opened data windows on top of the screen. It begins by referencing the *IApplicationWindows* interface. A collection of data windows is returned by applying the *DataWindows* method. Similar to enumerators, the *Reset* method sets the collection's sequence to the beginning. You can then apply the *Next* method to retrieve the data windows.

You can examine what type of data window you have by using the *TypeOf* operator. The preceding macro tests for the Magnification window to refresh its content after it has been moved. The Magnification window is an instance of the *MapInsetWindow* class. Use *IOverviewWindow* when searching for the Overview window.

The *IDataWindow* interface provides common properties and methods for both types of data windows. For instance, the *QueryPosition* and

PutPosition methods allow you to determine the position of the window and move it to a new location.

Using the Overview Window

Objects of the *OverviewWindow* class cannot be created. You need to use the *OverviewWindowFactory* co-class to create new Overview windows. The following procedure shows you how to create an Overview window and set the layer displayed in the window.

**CODE
VBA15-2**

```
Public Sub OpenOverviewWindow()
  ' Create the overview window.
  Dim pDataWindowFactory As IDataWindowFactory
  Dim pOverviewWindow As IOverviewWindow
  Set pDataWindowFactory = New OverviewWindowFactory
  If Not pDataWindowFactory.CanCreate(Application) Then
    MsgBox "Unable to create overview window."
    Exit Sub
  End If
  Set pOverviewWindow = pDataWindowFactory. _
  Create(Application)
  pOverviewWindow.Show True
  ' Set the layer displayed in the window.
  Dim pMxDocument As IMxDocument
  Dim pMap As IMap
  Dim pLayer As ILayer
  Dim pOverview As IOverview
  Set pMxDocument = Application.Document
  Set pMap = pMxDocument.FocusMap
  Set pLayer = pMap.Layer(0)
  Set pOverview = pOverviewWindow.Overview
  pOverview.OverlayGridLayer = pLayer
End Sub
```

In the preceding example, *DataWindowFactory* is used to create a new instance of the Overview window. The *CanCreate* method checks to make sure the Overview window can be created. The result of creating an Overview window is unpredictable when ArcMap is in layout mode. You should always check the ArcMap mode first, and change it if necessary.

When opening the Overview window, you should always set the layer that is displayed in the window. In the preceding procedure, the first layer of the activated map is shown in the window. The default symbol for showing the displayed extent is a red box. You can change it by setting *AoiFillSymbol* in the *IOverview* interface.

Using the Magnifier Window

The Magnifier window is modeled by the *MapInsetWindow* object in ArcMap. This window provides a magnified view of a portion of the map. The following procedure shows you how to create a magnification map and set its zoom factor to 500%.

**CODE
VBA15-3**

```
Public Sub OpenMapInsetWindow()
  ' Create a magnification window
  ' and set its zoom factor to 500%.
  Dim pDataWindowFactory As IDataWindowFactory
  Dim pMapInsetWindow As IMapInsetWindow
  Dim pMapInset As IMapInset
  Set pDataWindowFactory = New MapInsetWindowFactory
  If Not pDataWindowFactory.CanCreate(Application) Then
    MsgBox "Unable to create magnification window."
    Exit Sub
  End If
  Set pMapInsetWindow = pDataWindowFactory. _
  Create(Application)
  pMapInsetWindow.Show True
  Set pMapInset = pMapInsetWindow.MapInset
  pMapInset.ZoomPercent = 500
End Sub
```

Because *MapInsetWindow* is a part of a non-creatable class, you must use the corresponding window factory to create one. As shown in the preceding example, you should always test to make sure you can create a Map Inset window before applying the *Create* method. In certain situations, ArcMap cannot create the *MapInsetWindow* object.

Once the *MapInsetWindow* object is created, you can display it using the *Show* method. To change the zoom factor, you can reference the *IMapInset* interface through the *MapInset* property. Next, you use

ZoomPercent to set the zoom factor. You can use a negative value, which actually zooms out.

As part of your application, you may also want to close a specific Map Inset or Overview window. You can name the *Overview* or *MapInset* object and access them by name. The following code segment shows you how to name an *Overview* object.

```
pOverview.Name = "MyOVWin"
```

In the preceding example, *pOverview* is a reference to the *IOverview* interface. Earlier in this chapter, you saw how you could access the opened data windows. As you iterate through the data windows, you can check for the name to find the desired data window. The following example shows you how to check for the name and how to close the window.

```
If TypeOf pDataWindow Is IOverviewWindow Then
   Set pOverviewWindow = pDataWindow
   If pOverviewWindow.Overview.Name = "MyOVWin" Then
     pOverviewWindow.Show False
   End If
End If
```

In the preceding example, because the *Overview* method is available only through *IOverviewWindow*, a reference to this interface is obtained by querying the *IDataWindow* interface.

CHAPTER **16**

Using Page Layouts

YOU CAN USE THE PAGE LAYOUT OF ARCMAP to prepare a map for plotting. Although you can print from the map view, the page layout holds map surrounds such as legend and north arrow. The page layout has a virtual page that represents your hardcopy map. You arrange the map, map surrounds, and text elements on this page. Using ArcObjects, you can automate the process of arranging a hardcopy map. You can use one of the existing layout templates to quickly prepare a layout, or you can create and place all elements through your macro. This chapter teaches you both approaches through an application that can do both.

Defining the Application

The layout application presented in this chapter has three functions. It can generate a layout from a template, it can clear out the map surrounds from the page layout, and it can prepare a predefined layout by creating and placing the map surrounds. You will create three macros and several supporting procedures for this application. Each macro is a public VBA procedure that can be executed from a menu item or command button. You can also run a macro by selecting Tools | Macros | Macros.

The supporting procedures are private programs that are available only to the macros. Most of the functionality is provided through the private procedures, and the macros simply call these procedures as appropriate. This chapter concentrates on the programming code. It does not discuss how to create the user interface to run the macros. You learned how to do that in previous chapters.

Start by opening a new ArcMap document. Do not forget to save your document as you add the procedures. Create two data sets and name them *Detail* and *Overview*. The layout you will create uses the *Detail* data set for the main map frame, and the *Overview* data set for the overview map frame. Load some layers into the *Detail* data set. In addition, load one of the layers of the *Detail* data set into the *Overview* data set. The example presented here uses the *dtl_st* shape file from the CD-ROM number 3 of the ESRI data.

Next, open the VBA editor by selecting Tools | Macros | Visual Basic Editor. In the VBA editor, double click on the *ThisDocument* object in the Project window to open the Code window. In the empty Code window, enter the following two statements.

```
Option Explicit
Const c_dblMargin As Double = 0.5
```

Option Explicit ensures that you declare all variables and do not mistype a variable name in the procedure's body. The *Const* (constant) declaration sets the page margin in the layout to 0.5 inches.

Using Templates

A quick way to create a layout is to use existing templates. Add the following procedures to the Code window of the *ThisDocument* object in the VBA editor.

CODE
VBA16-1

```
Public Sub UseTemplate()
    ' Set layout using the template Wizard.
    Dim pMxDocument As IMxDocument
    Dim pChangeLayout As IChangeLayout
    Set pMxDocument = Application.Document
    Set pChangeLayout = pMxDocument
    pChangeLayout.ChangeLayout
End Sub
```

The preceding macro uses the *IChangeLayout* interface of the ArcMap document to display templates and apply the selected template. This action is similar to clicking on the Change Layout command button of the Layout toolbar.

Clearing the Layout

This macro removes all graphic and frame elements in the page layout except the map frames. Deleting the map frames removes the data sets. This macro is also used by the third macro, named *CreateLayout*, in this application. The third macro programmatically creates the page layout graphic and frame elements after it has cleared any existing ones.

The elements on the page layout include the frames, such as map or table frames, and graphics such as text or shape elements. To remove an element, the macro accesses the page layout's graphics container. The graphics container acts as a collection storing the elements of the page layout. Add the following procedure to the Code window.

CODE VBA16-2

```
Public Sub ClearPageLayout()
  ' Remove all frames except maps.
  Dim pMxDocument As IMxDocument
  Dim pPageLayout As IPageLayout
  Dim pGraphicsContainer As IGraphicsContainer
  Dim pElementsSet As ISet
  Dim pElement As IElement
  Dim pActiveView As IActiveView
  ' Access page layout.
  Set pMxDocument = Application.Document
  Set pPageLayout = pMxDocument.PageLayout
  Set pGraphicsContainer = pPageLayout
  Set pActiveView = pPageLayout
  ' First create a set of elements to be deleted.
  Set pElementsSet = New esriCore.Set
  pGraphicsContainer.Reset
  Set pElement = pGraphicsContainer.Next
  Do While Not pElement Is Nothing
    If Not TypeOf pElement Is IMapFrame Then
      pElementsSet.Add pElement
```

```
      End If
      Set pElement = pGraphicsContainer.Next
   Loop
   ' Next remove the elements from the
   ' graphics container.
   pElementsSet.Reset
   Set pElement = pElementsSet.Next
   Do While Not pElement Is Nothing
      pGraphicsContainer.DeleteElement pElement
      Set pElement = pElementsSet.Next
   Loop
   ' Finally refresh the view.
   pActiveView.Refresh
End Sub
```

The preceding macro references the graphics container by querying the *IPageLayout* interface. The *IGraphicsContainer* interface, similar to other collection objects, has the *Reset* and *Next* methods to iterate through its members. This interface also has the *DeleteAllElements* method, but you cannot use it because you do not want to remove the map frames. As you iterate through the members of the container, the macro references each member with an object of the *Element* abstract class.

The *Element* abstract class provides the *IElement* interface that is implemented by all elements that can be placed on the page layout. You use the *TypeOf* operator to determine which co-class is the element.

This macro does not delete the elements directly out of the graphics container, because this adversely affects the looping process. Instead, it creates another collection set using the *ISet* interface to hold the elements to be deleted. The macro then iterates through the new set and removes the elements from the graphics container. When you run this macro, it removes all graphics and frame elements except the *Detail* and *Overview* map frames.

Creating a Layout

The third macro in this application creates a predefined layout based on a user-defined page size. Because there are several tasks to complete, the macro breaks up its code into logical units of private procedures. The macro is a public procedure, as shown in the following code listing that calls the private procedures. Each private procedure is listed and explained in the sections that follow. Add the following public procedure to the Code window.

**CODE
VBA16-3**

```
Public Sub CreateLayout()
    ' This VBA macro gets the page
    ' size from the user and creates
    ' a layout. It requires two map
    ' frames named Detail and Overview.
    Dim pOverviewElement As IElement
    Dim pDetailElement As IElement
    '

    ' (1) Remove any existing elements on
    '     the page layout except map frames.
    ClearPageLayout
    '

    ' (2) Get the page size and orientation
    '     from the user and apply it to the
    '     page layout.
    If Not SetPageSize() Then
        Exit Sub
    End If
    '

    ' (3) Position the detail map frame.
    Set pDetailElement = PlaceDetailMapFrame
    If pDetailElement Is Nothing Then
        Exit Sub
    End If
    '

    ' (4) Position the overview map frame.
    Set pOverviewElement = PlaceOverviewMapFrame
    If pOverviewElement Is Nothing Then
        Exit Sub
    End If
```

```
                        '
                        ' (5) Add legend, north arrow and scalebar.
                        AddLegend pDetailElement
                        AddNorthArrow pDetailElement
                        AddScalebar pDetailElement
                        '
                        ' (6) Add two text areas for the
                        '    user to add title and subtitle.
                        AddText
                        ' Page layout is refreshed in the
                        ' AddText procedure.
                      End Sub
```

The first task in the macro is calling the *ClearPageLayout* macro to remove existing page layout elements. It then calls several procedures to carry out specific tasks, starting with setting the page size.

Setting the Page Size

The *CreateLayout* macro builds the same layout regardless of the page size. The user of your macro selects the page size and orientation from a list. This procedure displays the list of page sizes and orientation and applies these attributes based on user selection. The procedure returns a *True* Boolean value if successful. Add the following procedure to the Code window.

CODE
VBA16-4

```
Private Function SetPageSize() As Boolean
   ' Show the page size choices and
   ' set the page size based on the
   ' user's selection.
   Dim pMxDocument As IMxDocument
   Dim pPageLayout As IPageLayout
   Dim pPage As IPage
   Dim pListDialog As IListDialog
   Dim pCoordinateDialog As ICoordinateDialog
   Dim blnOK As Boolean
   Dim dblWidth As Double
   Dim dblHeight As Double
   Dim vResponse As Variant
   ' Access the page object.
```

```
Set pMxDocument = Application.Document
Set pPageLayout = pMxDocument.PageLayout
Set pPage = pPageLayout.Page
' List the page size and orientation
' choices.
Set pListDialog = New ListDialog
pListDialog.AddString "Letter - Portrait"
pListDialog.AddString "Letter - Landscape"
pListDialog.AddString "Legal - Portrait"
pListDialog.AddString "Legal - Landscape"
pListDialog.AddString "C - Portrait"
pListDialog.AddString "C - Landscape"
pListDialog.AddString "D - Portrait"
pListDialog.AddString "D - Landscape"
pListDialog.AddString "E - Portrait"
pListDialog.AddString "E - Landscape"
pListDialog.AddString "Custom Page Size"
blnOK = pListDialog.DoModal _
("Select a page size and orientation", 0, 0)
If Not blnOK Then
  SetPageSize = False
  Exit Function
End If
' Apply the selected page size
' and orientation.
Select Case pListDialog.Choice
  Case 0
    ' Letter - Portrait
    pPage.FormID = esriPageFormLetter
    pPage.Orientation = 1
  Case 1
    ' Letter - Landscape
    pPage.FormID = esriPageFormLetter
    pPage.Orientation = 2
  Case 2
    ' Legal - Portrait
    pPage.FormID = esriPageFormLegal
    pPage.Orientation = 1
  Case 3
    ' Legal - Landscape
```

```
      pPage.FormID = esriPageFormLegal
      pPage.Orientation = 2
Case 4
  ' C - Portrait
      pPage.FormID = esriPageFormC
      pPage.Orientation = 1
Case 5
  ' C - Landscape
      pPage.FormID = esriPageFormC
      pPage.Orientation = 2
Case 6
  ' D - Portrait
      pPage.FormID = esriPageFormD
      pPage.Orientation = 1
Case 7
  ' D - Landscape
      pPage.FormID = esriPageFormD
      pPage.Orientation = 2
Case 8
  ' E - Portrait
      pPage.FormID = esriPageFormE
      pPage.Orientation = 1
Case 9
  ' E - Landscape
      pPage.FormID = esriPageFormE
      pPage.Orientation = 2
Case 10
  ' Custom Page Size
  ' Get the page size from the user.
      Set pCoordinateDialog = New CoordinateDialog
      Do
        ' Use a loop in case the user
        ' has to re-enter the page size.
        blnOK = pCoordinateDialog.DoModal _
        ("Width & Height (inches)", 8.5, 11, 2, 0)
        If Not blnOK Then
          SetPageSize = False
          Exit Function
        End If
        dblWidth = pCoordinateDialog.X
```

```
            dblHeight = pCoordinateDialog.Y
            If dblWidth < 5 Or dblHeight < 5 Then
              vResponse = MsgBox _
              ("Page width or height cannot " & _
              "be less than 5 inches", _
              vbExclamation + vbOKCancel)
              If vResponse = vbCancel Then
                SetPageSize = False
                Exit Function
              End If
            Else
              Exit Do
            End If
        Loop
        pPage.Units = esriInches
        pPage.PutCustomSize pCoordinateDialog.X, _
        pCoordinateDialog.Y
    End Select
    ' Done
    SetPageSize = True
End Function
```

The *SetPageSize* function uses a List dialog to display a list of sizes and orientations. Based on the user's selection, the function applies the size and orientation to the layout's page. The layout's page object is accessed through the *Page* property of the page layout object. The value of the *Orientation* property of the *Page* object can be either 1 for portrait or 2 for landscape. The page size can be established in two different ways. You can set it to a standard size using the *FormID* property or you can use the *PutCustomize* method for custom sizes.

The preceding function uses the Coordinate dialog to get the width and height of the custom-size page. The function does not accept a custom size of less than 5 inches for the width or height. The custom-size input is inside a loop so that if the given sizes are too small the program can ask for new sizes after displaying an error message.

Positioning Map Frames

The *CreateLayout* macro expects two map frames to exist in the Arc-Map document. The map frames correspond to data sets in ArcMap's table of contents. The data sets must be named *Detail* and *Overview*. The macro searches for these data sets by name. The following two procedures show you how to find the map frame in the page layout and how to resize and position it.

Add the following procedure to the Code window for placing the map frame for the *Detail* data set. The function returns reference to the *IElement* interface of the map frame. You will need this interface later for associating legend, north arrow, and scale bar to the *Detail* map.

CODE
VBA16-5

```
Public Function PlaceDetailMapFrame() As IElement
    ' Position the existing detail map frame
    ' in the upper left corner.
    Dim pMxDocument As IMxDocument
    Dim pPageLayout As IPageLayout
    Dim pPage As IPage
    Dim pActiveView As IActiveView
    Dim dblPageWidth As Double
    Dim dblPageHeight As Double
    Dim dblMargin As Double
    Dim dblFrameWidth As Double
    Dim dblFrameHeight As Double
    Dim pElement As IElement
    Dim pEnvelope As IEnvelope
    Dim pMapFrame As IMapFrame
    Dim pTransform2d As ITransform2D
    Dim pPoint As IPoint
    Dim dblXMove As Double
    Dim dblYMove As Double
    ' Get the page size.
    Set pMxDocument = Application.Document
    Set pPageLayout = pMxDocument.PageLayout
    Set pActiveView = pPageLayout
    Set pPage = pPageLayout.Page
    pPage.QuerySize dblPageWidth, dblPageHeight
    dblMargin = c_dblMargin
```

```
' Calculate the map frame's new size.
' Frame width and height are the page width
' or height less the margin and less two
' inches for other frames.
dblFrameWidth = dblPageWidth - (2 * dblMargin + 2)
dblFrameHeight = dblPageHeight - (2 * dblMargin + 2)
' Access the map frame.
Set pElement = GetMapFrameByName("DETAIL")
If pElement Is Nothing Then
  MsgBox "Missing Detail map."
  Set PlaceDetailMapFrame = Nothing
  Exit Function
End If
' Get the map frame's current size.
Set pEnvelope = New Envelope
pElement.QueryBounds pActiveView.ScreenDisplay, _
pEnvelope
' Map frame should show the map
' at its current scale.
Set pMapFrame = pElement
pMapFrame.ExtentType = esriExtentDefault
' Resize and move the detail map frame.
Set pTransform2d = pMapFrame
Set pPoint = New Point
pPoint.X = pEnvelope.XMin
pPoint.Y = pEnvelope.YMax
With pTransform2d
  .Scale pPoint, dblFrameWidth / pEnvelope.Width, _
  dblFrameHeight / pEnvelope.Height
End With
dblXMove = dblMargin - pEnvelope.XMin
dblYMove = (dblPageHeight - dblMargin) - _
pEnvelope.YMax
pTransform2d.Move dblXMove, dblYMove
' Done
Set PlaceDetailMapFrame = pElement
End Function
```

The function starts by getting the page size through the *QuerySize* method of the *IPage* interface. It then calculates the desired map frame width and height using the page size and page margin values.

The *GetMapFrameByName* function finds and returns the map frame to the procedure. This function is described later in this section. Once you have the map frame, you can resize and move it using the *ITransform2d* interface of the map frame. First, you resize the map frame, holding the upper left corner in place. The ESRI documentation states that when resizing with the *Scale* method of *ITransform2d* it must be enclosed within a *With* block. The documentation does not offer any reason for it, and indeed applying the *Scale* method without a *With* block does not work.

The next procedure for placing the overview map is similar to the *PlaceDetailMapFrame* function. It has a few new statements for showing the *Detail* map's extent rectangle on the overview map. Add the following code to the Code window.

**CODE
VBA16-6**

```
Public Function PlaceOverviewMapFrame() As IElement
    ' Position the overview map frame
    ' in the upper right corner.
    Dim pMxDocument As IMxDocument
    Dim pPageLayout As IPageLayout
    Dim pPage As IPage
    Dim pActiveView As IActiveView
    Dim dblPageWidth As Double
    Dim dblPageHeight As Double
    Dim dblMargin As Double
    Dim dblFrameWidth As Double
    Dim dblFrameHeight As Double
    Dim pElement As IElement
    Dim pEnvelope As IEnvelope
    Dim pMapFrame As IMapFrame
    Dim pDetailMapFrame As IMapFrame
    Dim pTransform2d As ITransform2D
    Dim pPoint As IPoint
    Dim dblXMove As Double
    Dim dblYMove As Double
    Dim pLocatorRectangle As ILocatorRectangle
    Dim pBorder As IBorder
    Dim pSymbolBorder As ISymbolBorder
    Dim pLineSymbol As ILineSymbol
    Dim pRGBColor As IRgbColor
```

```
' Get the page size.
Set pMxDocument = Application.Document
Set pPageLayout = pMxDocument.PageLayout
Set pActiveView = pPageLayout
Set pPage = pPageLayout.Page
pPage.QuerySize dblPageWidth, dblPageHeight
dblMargin = c_dblMargin
' Overview map frame's new width
' and height are 2 inches.
dblFrameWidth = 2
dblFrameHeight = 2
' Access the overview map frame.
Set pElement = GetMapFrameByName("OVERVIEW")
If pElement Is Nothing Then
  MsgBox "Missing Overview map."
  Set PlaceOverviewMapFrame = Nothing
  Exit Function
End If
' Get the map frame's
' current size.
Set pEnvelope = New Envelope
pElement.QueryBounds pActiveView.ScreenDisplay, _
pEnvelope
Set pMapFrame = pElement
pMapFrame.ExtentType = esriExtentDefault
' Resize and move the overview map frame.
Set pTransform2d = pMapFrame
Set pPoint = New Point
pPoint.X = pEnvelope.XMin
pPoint.Y = pEnvelope.YMax
With pTransform2d
  .Scale pPoint, _
  dblFrameWidth / pEnvelope.Width, _
  dblFrameHeight / pEnvelope.Height
End With
dblXMove = dblPageWidth - _
(dblMargin + dblFrameWidth) - pEnvelope.XMin
dblYMove = (dblPageHeight - dblMargin) - _
pEnvelope.YMax
pTransform2d.Move dblXMove, dblYMove
```

```
' Add the extent box to the overview map.
' Change the extent rectangle
' color to red.
pMapFrame.RemoveAllLocatorRectangles
Set pLocatorRectangle = New LocatorRectangle
Set pDetailMapFrame = GetMapFrameByName("DETAIL")
Set pLocatorRectangle.MapFrame = pDetailMapFrame
Set pBorder = pLocatorRectangle.Border
Set pSymbolBorder = pBorder
Set pLineSymbol = pSymbolBorder.LineSymbol
Set pRGBColor = New RgbColor
pRGBColor.Red = 255
pRGBColor.Green = 0
pRGBColor.Blue = 0
pLineSymbol.Color = pRGBColor
pLineSymbol.Width = 1.5
pSymbolBorder.LineSymbol = pLineSymbol
pLocatorRectangle.Border = pSymbolBorder
pMapFrame.AddLocatorRectangle pLocatorRectangle
' Done
Set PlaceOverviewMapFrame = pElement
End Function
```

The rectangle that represents the extent of the *Detail* map on the overview map is known as the locator rectangle. The locator rectangle is added to the overview map at the end of the preceding procedure. The procedure associates the *MapFrame* property of the locator rectangle with the *Detail* map frame so that it always matches the extent of the *Detail* map. It also changes the border symbol to a red line so that it is more noticeable.

The two functions that position the *Detail* and *Overview* maps use a procedure named *GetMapFrameByName* to access the appropriate map frame. This procedure accepts the name of the desired map and returns a reference to the *IElement* interface of the map frame. Add the *GetMapFrameByName* function, as shown in the following, to the Code window.

```
Public Function GetMapFrameByName(strMapName) As IElement
   ' Return the map frame that
   ' holds a map (dataset) with
   ' the given name.
   Dim pMxDocument As IMxDocument
   Dim pPageLayout As IPageLayout
   Dim pGraphicsContainer As IGraphicsContainer
   Dim pElementsSet As ISet
   Dim pElement As IElement
   Dim pActiveView As IActiveView
   Dim pMapFrame As IMapFrame
   ' Access the page layout.
   Set GetMapFrameByName = Nothing
   Set pMxDocument = Application.Document
   Set pPageLayout = pMxDocument.PageLayout
   Set pGraphicsContainer = pPageLayout
   ' Loop through graphics elements
   ' and examine each one.
   pGraphicsContainer.Reset
   Set pElement = pGraphicsContainer.Next
   Do While Not pElement Is Nothing
      If TypeOf pElement Is IMapFrame Then
        Set pMapFrame = pElement
        If UCase(pMapFrame.Map.Name) = strMapName Then
           Set GetMapFrameByName = pElement
           Exit Do
        End If
      End If
      Set pElement = pGraphicsContainer.Next
   Loop

End Function
```

The preceding function accesses the graphics container object of the page layout and examines each element in the container. It selects and returns the element that is a type of map frame and has a map name that matches the function's input name.

Adding a Rectangle

The two map frames added to the layout in the previous section have their own borders. However, for the legend, north arrow, and text elements you need to draw a rectangle around them. The following procedure accepts a rectangle in the form of an *Envelope* object and draws it on the page layout. The procedure is used later by other procedures that place the legend, north arrow, and text elements. Add the following procedure to the Code window.

CODE
VBA16-8

```
Private Sub AddRectangle(pEnvelope As IEnvelope)
    ' Draw a rectangle using the
    ' given envelope.
    Dim pMxDocument As IMxDocument
    Dim pPageLayout As IPageLayout
    Dim pActiveView As IActiveView
    Dim pGraphicsContainer As IGraphicsContainer
    Dim pRectangleElement As IElement
    Dim pFillShapeElement As IFillShapeElement
    Dim pFillRGBColor As IRgbColor
    Dim pLineRGBColor As IRgbColor
    Dim pFillSymbol As IFillSymbol
    Dim pLineSymbol As ILineSymbol
    ' Access the page layout.
    Set pMxDocument = Application.Document
    Set pPageLayout = pMxDocument.PageLayout
    Set pActiveView = pPageLayout
    Set pGraphicsContainer = pPageLayout
    ' Set the fill color to be
    ' transparent.
    Set pFillRGBColor = New RgbColor
    pFillRGBColor.Transparency = 0
    Set pFillSymbol = New SimpleFillSymbol
    pFillSymbol.Color = pFillRGBColor
    ' Set the outline color to black.
    Set pLineRGBColor = New RgbColor
    pLineRGBColor.Red = 0
    pLineRGBColor.Green = 0
    pLineRGBColor.Blue = 0
    Set pLineSymbol = New SimpleLineSymbol
```

```
        pLineSymbol.Color = pLineRGBColor
        pLineSymbol.Width = 1#
        pFillSymbol.Outline = pLineSymbol
        ' Create the rectangle element.
        Set pRectangleElement = New RectangleElement
        Set pFillShapeElement = pRectangleElement
        pFillShapeElement.Symbol = pFillSymbol
        ' Position the rectangle and
        ' display it on the page layout.
        pRectangleElement.Geometry = pEnvelope
        pRectangleElement.Activate pActiveView.ScreenDisplay
        pGraphicsContainer.AddElement pRectangleElement, 0
End Sub
```

The preceding procedure builds a rectangle element and sets its geometry to the given envelope. It also sets the fill color to be transparent and outline color to be black. To show the rectangle on the page layout you must first activate it and then add it to the page layout's graphics container. The *Activate* method requires the target display. It prepares the screen display for output.

Adding the Legend

Legend, north arrow, and scale bar elements on a page layout are known as map surround elements. To add one of these elements you need to create the map surround, build a frame for it, and add the frame to the page layout. These steps are shown in the following procedure. Add this procedure, used by the *CreateLayout* macro, to the Code window.

CODE
VBA16-9

```
Private Sub AddLegend(pElement As IElement)
    ' Add a legend for the given
    ' map frame.
    Dim pMxDocument As IMxDocument
    Dim pPageLayout As IPageLayout
    Dim pPage As IPage
    Dim pActiveView As IActiveView
    Dim dblPageWidth As Double
    Dim dblPageHeight As Double
    Dim dblMargin As Double
```

```
Dim dblFrameWidth As Double
Dim dblFrameHeight As Double
Dim pMapFrame As IMapFrame
Dim pMapSurroundFrame As IMapSurroundFrame
Dim pLegendElement As IElement
Dim pEnvelope As IEnvelope
Dim pFrameElement As IFrameElement
Dim pGraphicsContainer As IGraphicsContainer
Dim pID As New UID
Dim pMapSurround As IMapSurround
Dim dblX As Double
Dim dblY As Double
' Get the page size.
Set pMxDocument = Application.Document
Set pPageLayout = pMxDocument.PageLayout
Set pActiveView = pPageLayout
Set pGraphicsContainer = pPageLayout
Set pPage = pPageLayout.Page
pPage.QuerySize dblPageWidth, dblPageHeight
dblMargin = c_dblMargin
' Frame width is 2 inches. Frame
' height is page height less the margins
' and 2 inches for overview map.
dblFrameWidth = 2
dblFrameHeight = dblPageHeight - (2 * dblMargin + 2)
' Create a legend map surround.
Set pMapFrame = pElement
pID.Value = "esriCore.Legend"
Set pMapSurround = New Legend
Set pMapSurroundFrame = pMapFrame. _
CreateSurroundFrame(pID, pMapSurround)
' Size and position the new legend frame.
Set pFrameElement = pMapSurroundFrame
Set pLegendElement = pFrameElement
Set pEnvelope = New Envelope
' X and Y of the frame's
' upper left corner
dblX = dblPageWidth - (dblMargin + dblFrameWidth)
dblY = dblPageHeight - (dblMargin + 2)
pEnvelope.PutCoords dblX, _
```

```
          (dblY - dblFrameHeight + 2), _
          (dblX + dblFrameWidth), dblY
          ' Draw a rectangle around the
          ' legend frame.
          AddRectangle pEnvelope
          ' scale down the rectangle to _
          ' place the legend inside the
          ' drawn rectangle.
          pEnvelope.Expand 0.9, 0.9, True
          pLegendElement.Geometry = pEnvelope
          ' Add the legend to the page layout.
          pLegendElement.Activate pActiveView.ScreenDisplay
          pGraphicsContainer.AddElement pLegendElement, 0
    End Sub
```

A legend frame on the page layout needs to be associated with a map frame. In the preceding procedure the map frame for the *Detail* map is an input as a reference to the *IElement* interface. The following statements in the procedure create the legend frame and associate it with the *Detail* map frame.

```
Set pMapFrame = pElement
pID.Value = "esriCore.Legend"
Set pMapSurround = New Legend
Set pMapSurroundFrame = pMapFrame. _
CreateSurroundFrame(pID, pMapSurround)
```

The *CreateSurroundFrame* method in the preceding code segment creates a legend frame for the *Detail* map frame. When you positioned the map frames in an earlier section, you accessed the existing map frame and resized and scaled it. Because the frame for your map's legend is new, you will set the correct size and position at the same time you create the frame. Depending on the number of layers in a map, the legend frame may shrink or expand from its defined geometry. Therefore, in the procedure you draw a rectangle that houses the legend. You then slightly shrink the rectangle as the geometry of the legend frame.

Adding the North Arrow

A north arrow element is added to the page layout in the same manner as the legend. Add the following procedure to the Code window for placing a north arrow on the page layout.

**CODE
VBA16-10**

```
Private Sub AddNorthArrow(pElement As IElement)
    ' Add a north arrow for the
    ' given map frame.
    Dim pMxDocument As IMxDocument
    Dim pPageLayout As IPageLayout
    Dim pPage As IPage
    Dim pActiveView As IActiveView
    Dim dblPageWidth As Double
    Dim dblPageHeight As Double
    Dim dblMargin As Double
    Dim pID As New UID
    Dim pMapSurround As IMapSurround
    Dim dblFrameWidth As Double
    Dim dblFrameHeight As Double
    Dim pMapFrame As IMapFrame
    Dim pMapSurroundFrame As IMapSurroundFrame
    Dim pNorthArrowElement As IElement
    Dim pEnvelope As IEnvelope
    Dim pFrameElement As IFrameElement
    Dim pGraphicsContainer As IGraphicsContainer
    Dim dblX As Double
    Dim dblY As Double
    ' Get the page size.
    Set pMxDocument = Application.Document
    Set pPageLayout = pMxDocument.PageLayout
    Set pActiveView = pPageLayout
    Set pGraphicsContainer = pPageLayout
    Set pPage = pPageLayout.Page
    pPage.QuerySize dblPageWidth, dblPageHeight
    dblMargin = c_dblMargin
    ' North arrow frame width and _
    ' height are 2 inches.
    dblFrameWidth = 2
    dblFrameHeight = 2
```

```
' Create a north arrow map surround.
Set pMapFrame = pElement
pID.Value = "esriCore.MarkerNorthArrow"
Set pMapSurround = New MarkerNorthArrow
Set pMapSurroundFrame = pMapFrame. _
CreateSurroundFrame(pID, pMapSurround)
' Size and position the north arrow frame.
Set pFrameElement = pMapSurroundFrame
Set pNorthArrowElement = pFrameElement
Set pEnvelope = New Envelope
' X and Y of the upper left corner.
dblX = dblPageWidth - (dblMargin + dblFrameWidth)
dblY = dblMargin + dblFrameHeight
pEnvelope.PutCoords dblX, (dblY - dblFrameHeight), _
(dblX + dblFrameWidth), dblY
' Draw a rectangle on the page layout.
AddRectangle pEnvelope
' scale down the rectangle to
' fit the north arrow.
pEnvelope.Expand 0.9, 0.9, True
pNorthArrowElement.Geometry = pEnvelope
' Draw the north arrow on the page layout.
pNorthArrowElement.Activate pActiveView.ScreenDisplay
pGraphicsContainer.AddElement pNorthArrowElement, 0
End Sub
```

The preceding procedure accepts the default symbol for the new north arrow marker. You can change the symbol using the *IMarkerNorthArrow* interface.

Adding the Scale Bar

A scale bar element is added to the page layout in the same manner as the legend or north arrow. Add the following procedure to the Code window to place a scale bar on the page layout. The scale bar is placed just beneath the north arrow.

**CODE
VBA16-11**

```
Private Sub AddScalebar(pElement As IElement)
    ' Add a scalebar for the given map frame.
    Dim pMxDocument As IMxDocument
```

```
Dim pPageLayout As IPageLayout
Dim pPage As IPage
Dim pActiveView As IActiveView
Dim dblPageWidth As Double
Dim dblPageHeight As Double
Dim dblMargin As Double
Dim pID As New UID
Dim dblFrameWidth As Double
Dim dblFrameHeight As Double
Dim pMapFrame As IMapFrame
Dim pMapSurroundFrame As IMapSurroundFrame
Dim pScalebarElement As IElement
Dim pEnvelope As IEnvelope
Dim pFrameElement As IFrameElement
Dim pGraphicsContainer As IGraphicsContainer
Dim dblX As Double
Dim dblY As Double
' Get the page size.
Set pMxDocument = Application.Document
Set pPageLayout = pMxDocument.PageLayout
Set pActiveView = pPageLayout
Set pGraphicsContainer = pPageLayout
Set pPage = pPageLayout.Page
pPage.QuerySize dblPageWidth, dblPageHeight
dblMargin = 0.5
' Frame's width & height.
dblFrameWidth = 2
dblFrameHeight = 0.3
' Create a scalebar map surround.
Set pMapFrame = pElement
pID.Value = "esriCore.Scalebar"
Dim pMapSurround As IMapSurround
Set pMapSurround = New AlternatingScaleBar
Set pMapSurroundFrame = pMapFrame. _
CreateSurroundFrame(pID, pMapSurround)
' Size and position the scalebar.
Set pFrameElement = pMapSurroundFrame
Set pScalebarElement = pFrameElement
Set pEnvelope = New Envelope
' X and Y of upper left corner
```

```
        dblX = dblPageWidth - (dblMargin + dblFrameWidth)
        dblY = dblMargin + dblFrameHeight
        pEnvelope.PutCoords dblX, (dblY - dblFrameHeight), _
        (dblX + dblFrameWidth), dblY
        ' Scale the rectangle to fit the
        ' scalebar inside the rectangle
        ' drawn for the north arrow.
        pEnvelope.Expand 0.9, 0.9, True
        pScalebarElement.Geometry = pEnvelope
        ' Add the scalebar to the page layout.
        pScalebarElement.Activate pActiveView.ScreenDisplay
        pGraphicsContainer.AddElement pScalebarElement, 0
End Sub
```

The following are available scale bar styles. In the preceding procedure you created an alternating scale bar.

- Alternating scale bar

- Double alternating scale bar

- Hollow scale bar

- Single-division scale bar

- Scale line

- Stepped scale line

- Scale text

Adding Text Elements

You add text elements to the page layout as graphic elements. The following procedure adds two text elements for the title and subtitle of the layout. Add this procedure to the Code window.

CODE VBA16-12

```
Private Sub AddText()
    ' Add to two text elements for
    ' title and subtitle; and refresh
    ' the page layout.
    Dim pMxDocument As IMxDocument
    Dim pPageLayout As IPageLayout
```

```
Dim pActiveView As IActiveView
Dim pGraphicsContainer As IGraphicsContainer
Dim pGraphicsContainerSelect As IGraphicsContainerSelect
Dim pPage As IPage
Dim dblPageWidth As Double
Dim dblPageHeight As Double
Dim dblMargin As Double
Dim dblFrameWidth As Double
Dim dblFrameHeight As Double
Dim pEnvelope As IEnvelope
Dim pPoint As IPoint
Dim dblX As Double
Dim dblY As Double
Dim pTextElement As ITextElement
Dim pElement As IElement
Dim pTextSymbol As ITextSymbol
' Get the page size.
Set pMxDocument = Application.Document
Set pPageLayout = pMxDocument.PageLayout
Set pActiveView = pPageLayout
Set pGraphicsContainer = pPageLayout
Set pGraphicsContainerSelect = pPageLayout
Set pPage = pPageLayout.Page
pPage.QuerySize dblPageWidth, dblPageHeight
dblMargin = c_dblMargin
' Frame width is the page width
' less the margins and 2 inches
' for other frames.
' Frame height is 2 inches.
dblFrameWidth = dblPageWidth - (2 * dblMargin + 2)
dblFrameHeight = 2
' X and Y of the upper left corner.
dblX = dblMargin
dblY = dblMargin + dblFrameHeight
' Draw a rectangle to house the
' two text elements.
Set pEnvelope = New Envelope
pEnvelope.PutCoords dblX, (dblY - dblFrameHeight), _
(dblX + dblFrameWidth), dblY
AddRectangle pEnvelope
' Create the title text element.
```

```
    Set pTextElement = New TextElement
    pTextElement.Text = "Double click here to add text"
    Set pTextSymbol = pTextElement.Symbol
    pTextSymbol.Size = 24
    pTextSymbol.HorizontalAlignment = esriTHACenter
    pTextElement.Symbol = pTextSymbol
    Set pElement = pTextElement
    Set pPoint = New Point
    pPoint.X = dblMargin + (dblFrameWidth / 2)
    pPoint.Y = dblMargin + (2 * dblFrameHeight / 3)
    ' Position the text title and add it
    ' to the page layout.
    pElement.Geometry = pPoint
    pGraphicsContainer.AddElement pTextElement, 0
    ' Create the subtitle text element.
    Set pTextElement = New TextElement
    pTextElement.Text = "Double click here to add text"
    Set pTextSymbol = pTextElement.Symbol
    pTextSymbol.Size = 14
    pTextSymbol.HorizontalAlignment = esriTHACenter
    pTextElement.Symbol = pTextSymbol
    Set pElement = pTextElement
    Set pPoint = New Point
    pPoint.X = dblMargin + (dblFrameWidth / 2)
    pPoint.Y = dblMargin + (dblFrameHeight / 3)
    ' Position the text title and add it
    ' to the page layout.
    pElement.Geometry = pPoint
    pGraphicsContainer.AddElement pTextElement, 0
    ' Done. Unselect all layout elements
    ' and refresh the page layout.
    pGraphicsContainerSelect.UnselectAllElements
    pActiveView.Refresh
End Sub
```

When you add text to the page layout, use a point as the geometry of the text element. The point becomes the insertion point for the text. In the preceding procedure, once the text element is created the text size is changed. To change the text size, you first access the text symbol. You then change the size and then use the *Symbol* object to reset the *Symbol* property of the text element. You can draw the text elements by simply adding them to the graphics container of the page layout. At the end of this procedure you deselect all elements on the page layout and refresh the view.

Testing the Application

After you have added all procedures in this chapter to an ArcMap document, you are ready to test the application. This chapter does not build the user interface; you learned how to do that in earlier parts of the book. You do not need the user interface to test your application. However, if you do plan to provide a user interface, make sure you test the application again with the user interface. Open the Macros dialog by selecting Tools | Macros | Macros. You should see the following three macros in the Macro dialog.

- *ClearPageLayout*

- *CreateLayout*

- *UseTemplate*

Select one of the macros and click on the Run button. Figure 16-1 shows the result of the *CreateLayout* macro with the *dtl_st* shape file using legal-size page and landscape orientation.

Fig. 16-1. Result of the CreateLayout macro.

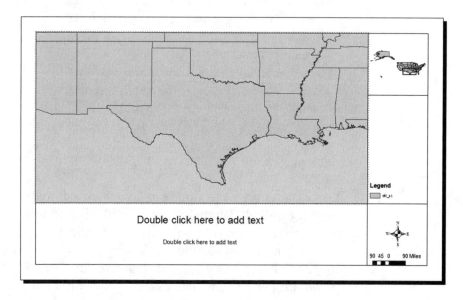

Index